# INTELLIGENT BROADBAND NETWORKS

# INTELLIGENT BROADBAND NETWORKS

*Edited by*

**Iakovos Venieris**
*National Technical University of Athens, Greece*

and

**Heinrich Hussmann**
*Dresden University of Technology, Germany*

JOHN WILEY & SONS
Chichester • New York • Weinheim • Brisbane • Singapore • Toronto

Copyright ©1998 by    John Wiley & Sons Ltd
Baffins Lane, Chichester,
West Sussex, PO19 1UD, England

*National*          01243 779777
*International*    (+44) 1243 779777

e-mail (for orders and customer service enquiries): cs-books@wiley.co.uk

Visit our Home Page on http://www.wiley.co.uk or http://www.wiley.com

*Other Wiley Editorial Offices*

John Wiley & Sons, Inc., 605 Third Avenue,
New York, NY 10158-0012, USA

WILEY-VCH Verlag GmbH, Pappelallee 3,
D-69469 Weinheim, Germany

Jacaranda Wiley Ltd, 33 Park Road, Milton,
Queensland 4064, Australia

John Wiley & Sons (Asia) Pte Ltd, 2 Clementi Loop #02-01,
Jin Xing Distripark, Singapore 129809

John Wiley & Sons (Canada) Ltd, 22 Worcester Road,
Rexdale, Ontario, M9W 1L1, Canada

**British Library Cataloguing in Publication Data**

A catalogue record for this book is available from the British Library

ISBN 0-471-98094-3

Produced from camera-ready copy supplied by the Editors
Printed and bound in Great Britain by Biddles Ltd, Guildford and King's Lynn
This book is printed on acid-free paper responsibly manufactured from sustainable forestry in which at least two trees are planted for each one used for paper production.

# CONTENTS

# FOREWORD

I was most happy to accept the invitation of Prof. Iakovos Venieris, Prof. Heinrich Hussmann, editors of this book, and Mr. Bert Koch, manager of the INSIGNIA Project, to write a foreword for the book in hand which comprises the common efforts of a number of researchers from different European countries made possible by the European Union ACTS (Advanced Communications Technologies and Services) Programme.

The objective of the ACTS Programme is the development and early deployment of advanced communication systems and services in Europe, enabling effective competition in global markets. Thus ACTS has moved one step beyond from its precedent programme RACE by putting its emphasis on the implementation of advanced networks and services demonstrated through a number of transnational trials, rather than on the detailed specification and design of standalone systems. In this respect ACTS aims at establishing the basic framework for the research and development of advanced information services through the experience and expertise gained by the balanced collaboration of scientists from research institutions, universities, manufacturers, public operators and service providers.

Specifically ACTS has addressed the areas of Interactive Digital Multimedia Services, Photonic Technologies, High-Speed Networking, Mobility and Personal Communication Networks, Intelligence in Networks and Services, Quality, Security and Safety of Communication Systems and Services, complemented by a set of Horizontal Actions aiming at development of consensus and concentration of activities performed within the ACTS framework beyond the EU borders.

As the acronym decided for the project indicates (IN and B-ISDN Signaling Integration on ATM platforms), INSIGNIA adopts a pragmatic evolutionary approach for the support of multimedia services that attempts to combine current developments around Broadband signaling and Intelligent Network Technology. To realize this not so straightforward combination a number of advanced developments enhancing the functionality of existing systems has been performed. The book does not only present these developments but also provides preceptive guidelines on how existing possibly dissimilar systems can interoperate in the context of an integrated service provision environment. The INSIGNIA project has achieved to demonstrate the validity of its concept by performing two transnational trials open to the public, a fact which further justifies the success of the project and contributes to the accomplishment of the ACTS programme objectives.

The INSIGNIA consortium brought together researchers from 14 organizations in eight European countries. It comprised five equipment manufacturers, three network operators, four research institutes and two universities. The cooperation within this well-balanced consortium led to a new perspective on Broadband communications which would have been impossible to achieve for any single involved organization. All partners of the consortium have contributed to a truly European contribution to the world-wide scientific community and to international standardization.

**Dr Spyros Konidaris**
*European Commission*

# PREFACE

There is a revolution going on in the field of telecommunication network architecture. It is now technically possible that traditional telecommunication services such as the telephone system are completely integrated into a computer network like the Internet. On the other hand, it is technically possible that a modern Broadband-ISDN provides multimedia services of a quality and reliability which is far beyond the reach of the current Internet. The essence of the revolution is a convergence of formerly unrelated areas of technology, in particular data network technology, software technology and traditional telecommunication technology.

As always during the course of a revolution, a great deal of confusion is created. Several parties are formed which compete for the ultimate way to combine the technologies, and strategists try to predict or influence the decision which of the parties will win. Unfortunately, this often leads into a situation where strategic arguments outweigh a detailed technical discussion.

This book addresses a central issue of the technology revolution, which is the architecture of a future network integrating traditional telecommunication services and modern multimedia services. Nevertheless, it takes a perspective which is different from most of the current heated discussion. The approach used here is conservative to some extent, since it takes into account the fact that large and powerful networks constitute a very high investment for the network operators which has to pay back. Therefore, it is investigated how the revolution could be replaced by an evolutionary development. The evolution builds on top of modern telecommunication technology but integrates concepts and features of modern data communication and software architectures. The telecommunication technology is represented here by the Broadband-ISDN, which is an advanced public ATM network, and by the 'Intelligent Network' (IN) architecture which is currently used for the provision of value-adding services in the telephone network. As a next step of evolution of this technology IN and B-ISDN should be combined. The central assumption of this book is that a combination of B-ISDN and IN can be achieved in a way which takes into account multimedia services and modern object-oriented software architecture. So it is not only a straightforward combination of IN and

B-ISDN which is described here but a carefully designed amalgamation of telecommunication and computing concepts.

This book is the outcome of the international project INSIGNIA which was sponsored by the European Union as part of the ACTS research programme (Advanced Communications Technologies and Services). This project was a unique opportunity for a group of scientists and developers from several industrial companies and research institutions to go to a significant level of detail in the investigation of the outlined evolutionary approach. The result, as documented in this book, is a consistent set of architecture definition, protocol enhancements and service design methodology which has proven its practical viability by several prototype implementations and a large trial in the context of an international high-speed network.

Although following a clear evolutionary approach, the project did not ignore the trends which are visible from other more revolutionary ways to define the architecture of future networks. An analysis has been carried out how a synergy can be achieved between the INSIGNIA approach and two noticeable other approaches, which are the architecture proposed by the TINA consortium and the achievements of the Internet community. The results of this analysis are documented in this book as well.

This book is a common product of a large group of authors working in several countries of Europe. There were only three physical meetings of all the authors, and none with the publisher. This has been made possible by using the best telecommunication infrastructure currently available. The project members hope that their work contributes to a future European communication infrastructure that will be even more powerful and user-friendly, and therefore will enable even more efficient collaborative work of international teams.

## ACKNOWLEDGEMENTS

The editors would like to thank all members of the INSIGNIA project for their excellent work. In particular, we would like to thank those colleagues who have contributed to this book significantly but who are not mentioned as authors:
D. Blaiotta, M. Breda, P.F. Chimento, S. Corti, S. Daneluzzi, D. Fava, M. Grech, M. Listanti, M. Lombari, G.K. Mamais, A. E. Milner, I.G.M.M. Niemegeers, G.A. Politis, E.N. Protonotarios, L. Ronchetti, K.-U. Stein, B.J. van Beijnum, M. Varisco, L. Vezzoli, G. von der Straten.

Thanks go also to Christian Rauscher and his team in Chichester for the effective and friendly guidance they provided for this work.

Any figures credited to the ITU have been reproduced with permission of the copyright holder and the complete volume (s) from which the figures/texts reproduced are extracted can be obtained from:
International Telecommunication Union
Place des Nations-CH 1211 Geneva 20 (Switzerland)

Telephone: + 41 22 730 61 41 (English) / + 41 22 730 61 42 (France)
Telex: 421 000 uit ch / Fax: + 41 22 730 51 94
X.400: S=Sales; P=itu; C=ch
Internet: Sales@itu.int

**Iakovos S. Venieris**
*National Technical University of Athens*


**Heinrich Hussmann**
*Dresden University of Technology*

# CONTRIBUTORS

**F Bernabei**
*Fondazione Ugo Bordoni*
*Rome*

**H Brandt**
*GMD FOKUS*
*Berlin*

**G Chierchia**
*Fondazione Ugo Bordoni*
*Rome*

**F Cuomo**
*University of Rome "La Sapienza"*
*Rome*

**G De Zen**
*ITALTEL*
*Milan*

**L Faglia**
*ITALTEL*
*Milano*

**M Geipl**
*Deutsche Telecom AG*
*Darmstadt*

**L Gratta**
*Fondazione Ugo Bordoni*
*Rome*

**J Humphrey**
*GPT Ltd*
*Poole*

**H Hussmann**
*Dresden University of Technology*
*Dresden*

**G Karagiannis**
*University of Twente*
*Enschede*

**B Koch**
*Siemens AG*
*Munich*

**G Kolyvas**
*National Technical University of Athens*
*Athens*

**G Marino**
*ITALTEL*
*Milan*

**E Menduina Martin**
*TELEFONICA I+D*
*Madrid*

**V Nicola**
*University of Twente*
*Enschede*

**Ch Patrikakis**
*National Technical University of Athens*
*Athens*

**S Polykalas**
*National Technical University of Athens*
*Athens*

**G Prezerakos**
*National Technical University of Athens*
*Athens*

**S Salsano**
*University of Rome "La Sapienza"*
*Rome*

**P Todorova**
*GMD FOKUS*
*Berlin*

**A van der Vekens**
*Siemens AG*
*Munich*

**I S Venieris**
*National Technical University of Athens*
*Athens*

**F Zizza**
*ITALTEL*
*Milan*

# ACRONYMS AND ABBREVIATIONS

**AAL**   ATM Adaptation Layer
**ACTS**   Advanced Communications Technologies and Services
**AE**   Application Entity
**ALS**   Application Layer Structure
**AM**   Access Manager
**AMX**   ATM Multiplexer
**AoO**   Area of Origin
**AP**   Application Process
**API**   Application Programming Interface
**ARP**   Address Resolution Protocol
**ASE**   Application Service Element
**ASN.1**   Abstract Syntax Notation
**ASN**   ATM Switching Network
**ATM**   Asynchronous Transfer Mode
**ATMARP**   ATM Address Resolution Protocol
**AUs**   Adaptation Units
**B-**   Broadband
**BCC**   Bearer Connection Control
**BCM**   Basic Call Manager
**BCP**   Basic Call Process
**BCSM**   Basic Call State Model
**BCUSM**   Basic Call Unrelated State Model
**BER**   Basic Encoding Rules
**BNCM**   Basic Non Call Manager
**BRI**   Basic Rate Interface
**B-VC**   Broadband Video Conference
**CBR**   Constant Bit Rate
**CC**   (1) Call Control, (2) Call Configuration (Subsection 1.2.4.2, in the context of CS-2 IN-SSM), (3) Connection Coordinator (Chapter 3.2, in the context of TINA)
**CCF**   Call Control Function
**CCAF**   Call Control Agent Function
**CCM**   Call Control Manager

**CdPN**   Called Party Number
**CD-ROM**   Compact Disk-Read Only Memory
**CE**   Computing Element
**CEC**   Commission of the European Community
**CG**   Call Gap
**CgPN**   Calling Party Number
**CHILL**   CCITT High Level Language
**CID**   Conference Identifier
**CIDB**   Conference Information Data Base
**CIR**   Conference Information Record
**CM**   (1) Call Manager, (2) Connection Management (Chapter 3.2, in the context of TINA)
**CMIP**   Common Management Information Protocol
**CO**   Computational Objects
**COBI**   Connection Oriented Bearer Independent
**CORBA**   Common Object Request Broker Architecture
**CP**   Connection Performers
**CPP**   Centralized Point to Point
**CPE**   Customer Premises Equipment
**CPMP**   Centralized Point to Multi-point
**CS**   Capability Set
**CSCW**   Computer Supported Co-operative Work
**CTM**   Cordless Terminal Mobility
**CUC**   Call Unrelated Control
**CUSF**   Call Unrelated Service Function
**CUUI**   Call Unrelated User Interaction
**CV**   Connection View
**DAVIC**   Digital AudioVIsual Council
**DC**   EnD Control
**DCA**   EnD Control Agent
**DP**   (1) Detection Point, (2) Data Part (Section 1.2.6, in the context of SRF)
**DPE**   Distributed Processing Environment
**DPMP**   Distributed Point to Point
**DPP**   Distributed Point to Point
**DSL**   Distributed Service Logic
**DSS1**   Digital Subscriber Signaling System No. 1
**DTMF**   Dual Tone Multi Frequency
**EC**   Edge Control
**EDP**   Event Detection Point
**ES**   Is for Em-space which is inserted before the definition
**ESIOP**   Environment Specific Inter-ORB protocol
**FE**   Functional Entity
**FEA**   Functional Entity Action
**FEAM**   Functional Entity Access Manager
**FIM**   Feature Interaction Manager
**FRL**   Functional Routine Library

**FRM** Functional Routine Manager
**FSM** Finite State Machine
**GFM** Global Functional Model
**GFP** Generic Functional Protocol
**GIOP** General Inter-ORB Protocol
**GPE** Group Processor type E
**GSBB** Generic Service Building Block
**GSEP** Generic Session End Point
**GSI** Generic Service Interface
**GSL** Global Service Logic
**GUI** Graphical User Interface
**HTML** Hyper-Text Markup Language
**IA** Initial Agent
**IDL** Interface Definition Language
**IE** Information Elements
**IF** Information Flow
**IIOP** Internet Inter-ORB Protocol
**IM** Interface Mapper
**IMR** Interactive Multimedia Retrieval
**IMT-2000** International Mobile Telecommunication 2000
**IN** Intelligent Network
**IN-** Intelligent Network-
**INAP** IN Application Protocol
**INSIGNIA** IN and B-ISDN Signaling Integration over ATM platforms
**IP** Intelligent Peripheral
**IPA** B-IP Administrator
**ISDN** Integrated Services Digital Network
**ISODE** ISO Development Environment
**ISUP** ISDN User Part
**ITU-T** International Telecommunications Union
**IW** Interworking
**IWF** Interworking Function
**KOD** Karaoke on Demand
**KTN** Kernel Transport Network
**LAN** Local Area Network
**LAP-D** Link Access Procedure on the D channel
**LC** Link Control
**LCA** Link Control Agent
**LCP** Logical Connection Graph
**LEX** Local Exchange
**LM** Local Manager
**LN** Layer Network
**LNC** Layer Network Interface
**LPU** LAP-D Processor Unit
**M** Mandatory
**MBONE** Multicast backBONE network

**MC**   Maintenance Control
**MCU**   Multi-point Control Unit
**MOD**   Music On Demand
**MPEG**   Moving Pictures Experts Group
**MTP**   Message Transfer Part
**NAP**   Network Access Point
**NCM**   Non Call Manager
**NH**   National Host
**NIFM**   Non-IN Feature Manager
**NNI**   Network Node Interface
**NoV**   News on Video
**NRA**   Network Resource Architecture
**NRD**   Network-wide Resource Data
**NSM**   Non Switching Manager
**NWTTP**   Network Trail Terminator Points
**O**   Optional
**OAM**   Operation, Administration and Maintainance
**OCCRUI**   Out of Channel Call Related User Interaction
**OCCUUI**   Out Channel Call unrelated User Interaction
**ODL**   Object Definition Language
**ODP**   Open Distributed Processing
**ODR**   Origin Dependent Routing
**OMG**   Object Management Group
**OMT**   Object Modeling Technique
**ORB**   Object Request Broker
**OSF**   Open Systems Foundation
**OSI**   Open Systems Interconnection
**PA**   Provider Agent
**PC**   Personal Computer
**PCE**   Personal Computer type E
**PCM**   Party Control Manager
**PCP**   Physical Connection Graph
**PE**   Physical Entity
**PIA**   Point In Association
**PIC**   Point In Call
**PMU**   Processing and Memory Unit
**POI**   Point Of Initiation
**PON**   Passive Optical Network
**PoP**   Point of Presence
**POR**   Point Of Return
**PP**   Physical Plane
**PRI**   Primary Rate Interface
**PSTN**   Public Switched Telephone Network
**RCP**   Resource Control Part
**RFC**   Request for Comments
**RFP**   Resource Function Part

**RMR** Resource Management and Routing
**ROSE** Remote Operations Service Element
**RSVP** Resource ReSerVation Protocol
**QoS** Quality of Service
**SA** Service Architecture
**SAAL** Signaling ATM Adaptation Layer
**SACF** Single Associated Control Function
**SAO** Single Association Object
**SCs** Service Components
**SCCP** Signaling Connection Control Part
**SCF** Service Control Function
**SCF-DAM** SCF- Data Access Manager
**SCE** Service Creation Environment
**SCEF** Service Creation Environment Function
**SCME** SCF Management Entity
**SCP** Service Control Point
**SCS** Signaling Capability Sets
**SCSM** SCF Call State Model
**SCUAF** Service Control User Agent Function
**SDF** Service Data Function
**SDF-DM** SDF-Data Manager
**SDH** Synchronous Digital Hierarchy
**SDL** Specification and Description Language
**SDLC** Selective Discard Local Control
**SDOD** Service Data Object Directory
**SDP** Service Data Point
**SDT** SDL Design Tool
**SF** Service Feature
**SGI** Service Session Graph Interface
**SIB** Service Independent Building Block
**SLAC** SLEE Access Controller
**SLEE** Service Logic Execution Environment
**SLEM** Service Logic Execution Manager
**SLIDB** Service Logic and Instance Database
**SLL** Service Logic Library
**SLMB** Subscriber Line Module Broadband
**SLP** Service Logic Program
**SLPI** Service Logic Program Instance
**SLPL** Service Logic Program Library
**SLSIM** Service Logic Selection Interaction Manager
**SM** Switching Manager
**SMAF** Service Management Agent Function
**SME** Service Management Environment
**SMF** Service Management Function
**SN** Service Node
**SNMP** Simple Network Management Protocol

**SRF**   Specialized Resource Function
**SRSM**   SRF State Model
**SS7**   Signaling System No. 7
**SSC**   Service Support Component
**SSCF**   Service Specific Coordination Function
**SSCOP**   Service Specific Connection Oriented Protocol
**SSCP**   Service Switching and Control Point
**SSF**   Service Switching Function
**SSG**   Service Session Graph
**SSM**   (1) Switching State Model, (2) Service Session Manager (Chapter 3.2, in the context of TINA)
**SSO**   Service Support Object
**SSP**   Service Switching Point
**SSSM**   SCF Session State Model
**STB**  Set Top Box
**STM**   Sequence Transfer Mode
**SUI**   Simple User Interface
**SWC**   Switch Control
**TAXI**   Transparent Asynchronous transmitter/receiver Interfaces
**TC**   Transaction Capability
**TCAP**   Transaction Capability Application Part
**TCon**   Terminal Connection
**TCP**   Transmission Control Protocol
**TCP/IP**   TCP/Internet Protocol
**TCSM**   Terminal Communication Session Manager
**TDP**   Trigger Detection Point
**TDR**   Time Dependent Routing
**TE**   Terminal Equipment
**TEX**   Transit Exchange
**TINA**   Telecommunications Information Networking Architecture
**TINA-C**   TINA-Consortium
**TLA**   Terminal Layer Adapter
**TMB**   Trunk Module Broadband
**TMN**   Telecommunication Management Network
**ToD**   Time of Day
**ToW**   Day of Week
**UAP**   User Application
**UMTS**   Universal Mobile Telecommunications System
**UIS**   User Interaction Scripts
**UNI**   User Network Interface
**UPT**   Universal Personal Telecommunication
**USI**   User Service Information
**USM**   User Session Manager
**VBR**   Variable Bit Rate
**VC**   Virtual Channel
**VCI**   Virtual Channel Identifier

**VCR**   Video Cassette Recorder
**VI**   Virtual Intranet
**VoD**   Video on Demand
**VP**   Virtual Path
**VPI**   Virtual Path Identifier
**VPN**   Virtual Private Network
**VS**   Video Server

# Part 1

## INTELLIGENT NETWORKS AND BROADBAND SIGNALING FOR MULTIMEDIA SERVICES

# Chapter 1.1

# MULTIMEDIA SERVICES AND NETWORK CONCEPTS

Towards the end of the 20th century, a new technology starts to arise which brings together the joined force of several other technology waves. These other waves, which are affecting our daily life already, are affordable and decentralized computing (PCs), multimedia presentation techniques (CD-ROM) and a universal data network accessible for almost everybody (Internet). The topic of this book is a special technology which can help to integrate these separate trends in such a way that a unique new information and communication infrastructure evolves.

This book describes a concept for a network which enables communication services which will be as easy to use as the World Wide Web and which will provide a quality and intensity of presentation as achieved by CD-ROM today. The specific approach of this book is to describe how a homogeneous evolution of existing telecommunication networks into the new infrastructure can happen.

The main problem a network designer has to face nowadays is to select an efficient way for the rapid introduction of multimedia services to existing networks while guaranteeing the robustness of the adopted solution in view of the continuous progress in telecommunications technology. To achieve a solution that will provide safety and efficiency in future networks, the designer should start from a consideration of the requirements which are imposed to any network by multimedia services. On this background, the type and features of the available telecommunication infrastructure are to be examined. This will allow a straightforward definition of the modifications and enhancements required to a system for the support of multimedia services. When defining modifications and enhancements, it has always to be kept in mind, however, that the implementation should be open and flexible enough to incorporate future developments.

This book focuses on high-quality multimedia services delivered through a public telecommunication network. True multimedia communication almost always requires the transmission of one or more video channels, and therefore it has relatively high bandwidth requirements. Even when modern compression techniques are used, a bandwidth of significantly more than 64 Kbit/second is necessary to achieve a reasonable quality [1]. For this book, a public Broadband network is assumed which is able to provide scaleable bandwidth.

A multimedia service has to support several different media types simultaneously, where each medium has its own transmission characteristics. This results in a number of other capabilities which are required from the network [2]. The network has to be able to support *individual connections* with

*Intelligent Broadband Networks.* Edited by I. S. Venieris, H. Hussmann
© 1998 John Wiley & Sons Ltd.

*defined Quality of Service (QoS)*. The network has to provide a *connection management* facility which bundles together an arbitrary combination of traffic into a single call or single session. Various *communication configurations* need to be supported, in particular point-to-point, point-to-multipoint, multicast and broadcast communication. Moreover, the network has to cope for *synchronization* of different coupled media streams.

This introductory chapter gives the background and motivation for the detailed technical approach which is presented in this book. As a first step, the new communication services which will become available for the end user of the new infrastructure are discussed in Sections 1.1.2-1.1.3 below. In Section 1.1.4 multimedia service characteristics are analyzed to reveal the required functionality of the network at the user and control plane. The user plane includes all those network functions associated with the flow of the voice, video and data components of a user application. The control plane includes all network functions, protocols and messages required to manipulate the exchange of user information constituting an application. Currently operating telecommunication networks adopting the Integrated Services Digital Network (ISDN) and Internet technology as well as upcoming Broadband networks based on the Asynchronous Transfer Mode (ATM) technology are reviewed in Section 1.1.5 and their maturity in meeting the multimedia service demands is evaluated. The Intelligent Network concept for the support of advanced services is compared in Section 1.1.6 to other emerging alternatives with respect to the current state of the infrastructure including signaling system capabilities and the cost brought about by the network functionality enhancements. Finally, Section 1.1.7 introduces the motivation for the specific approach which is the topic of the remainder of this book.

## 1.1.1    Multimedia Applications and Services

The basic idea of multimedia technology is to use the natural way how human beings communicate. It is not natural for humans to restrict their attention to a single communication channel with a single presentation of information. For instance, two people involved in an intense discussion tend to use many ways to present information which is available. They speak, they make gestures, they make drawings, they retrieve information from files or books to show it to the other partner. In comparison to this natural way of communication, a telephone conversation is clearly limited, since it is restricted to a single communication channel with a single way to represent information. The telephone conversation uses a single medium, whereas the natural way to communicate is over multiple media. Based on these observations, the following definition makes sense:

> *Multimedia application*: A Multimedia Application is an application that requests the handling of two or more representation media (information types) simultaneously [3].

This definition is taken literally from a recommendation (F.700) of the International Telecommunications Union (ITU-T). The fact that the ITU-T work on these issues already gives an indication for the close coupling between telecommunication and multimedia applications. Currently the only available true multimedia applications (which are often distributed on CD-ROM) are application programs running on a standalone PC. However, since the whole purpose of multimedia technology is to make communication easier and simpler, it is absolutely natural for multimedia applications to be combined with telecommunication. So the logical next step is to proceed from stand-alone applications towards telecommunication services. Whenever we use the term *service* in this book, the implicit meaning is always *telecommunication service*. This is valid for the definition of a multimedia service which is again taken from ITU-T recommendation F.700:

> *Multimedia service*: Multimedia services are services that handle several types of media in a synchronized way from the user's point of view. A multimedia service may involve multiple parties, multiple connections, and the addition or deletion of resources and users within a single communication session [3].

One remark on a more grammatical level may also be helpful. Grammatically, the term 'multimedia' is an adjective and therefore it is advisable to always attach it to a noun, as in 'multimedia service'. Nevertheless, sometimes the term 'multimedia' is used as if it was a noun in itself.

## 1.1.2   Classification of Multimedia Services

This book deals with multimedia services delivered over public telecommunication networks. For these kind of services, a useful classification into sub-categories has been defined in [4] which is shown in Figure 1.1.1.

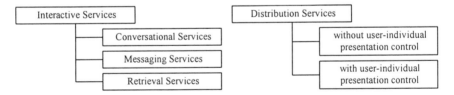

**Figure 1.1.1**
Classification of multimedia Broadband services.
Reproduced with permission of ITU from recommendation I.211 full source of original publication.

In the following, each of these service classes is briefly discussed and some examples are given. More exhaustive lists of examples can be found in [1, 5].

### 1.1.2.1 Interactive Conversational Services

This class of services is similar to the classical telephony service since it offers a means of communication for parties connected to the network. The content transferred by this service is mainly generated by the participating parties. Since, we are considering multimedia services here, the parties communicate over several synchronized media simultaneously. An arbitrary combination of at least two of the following media may be used:

- Audio conversation;
- Video conversation;
- Common access to computer applications (application sharing);
- Common access to shared space to write and draw (whiteboard);
- File transfer;
- Common access to media libraries (slide shows, video presentations).

In such a multimedia communication session, several different configurations of the involved parties can be required, dependent on the application area. The simplest case is a two-party bidirectional 'multimedia telephony'. A more complex case is a 'virtual meeting' with many participants, which requires a number of additional administrative services, like chairpersonship, voting procedures and others. In the extreme case, a large number of parties can be involved, e.g. in a lecture given over the network. However, in this case, much of the communication has to be unidirectional, showing a smooth transition into the class of distribution services.

There are various names used for services of the kind described above. In a business context, the term Computer Supported Co-operative Work (CSCW) has been established. However, also many other names appear which are derived from the application context. For instance, a multimedia among physicians with shared access to results of medical examination usually is called a 'telemedicine' application.

### 1.1.2.2 Interactive Messaging Services

Whereas the class of conversational services assumes an 'on line' communication where parties can react more or less in real time to actions of other parties, the class of messaging services provides a similar service in an 'off-line' style. In its simplest form, such a service can be realized by electronic mail using multimedia documents as message attachments. In a true multimedia messaging service, the message contains several media types in such a way that synchronization among the various media is given. An example is spoken commentary to text orgraphics which is recorded and played back in synchrony with pointing to parts of the document.

### 1.1.2.3 Interactive Retrieval Services

In an interactive retrieval service, the user gains access to multimedia information stored in information archives. There are two different phases of such a service. First, the user navigates to the required content, and afterwards, the content is presented to the user under interactive control by the user.

The most famous example of such a service is Video on Demand, where a telecommunication infrastructure is used to select and playback video information stored on large servers. This service was highly discussed as a potential competitor to normal TV broadcasting, but during the last few years it has turned out that an economic way to provide such a service is still far in the future. However, there are many other applications of the same kind of service in more specialized areas. For example, companies may use archives of training videos, and some universities already start to offer specialized courses in this form.

Although Video on Demand is somehow dominant in this service class, it should be kept in mind that true interactivity in multimedia retrieval can achieve much more than just control of video playback. For instance, in a computer-based training application, the user can be guided individually through a mesh of interrelated media pieces (video clips, interactive exercises etc.).

### 1.1.2.4 Distribution Services

According to the definitions used here, the distribution of a TV program over a data network may also be considered a multimedia service. It is not yet clear whether this kind of service has any chance to compete with other means of TV distribution, e.g. via satellite. However, there is the theoretical advantage that a TV distribution service based on a wide-area intelligent Broadband network can enable access to virtually any program available anywhere on the network, without any restriction in number of available programs. If the currently visible trend of convergence of networks (communication networks, data networks, broadcast networks) continues, then we may very well end up with a situation where a universal Broadband network will be based on satellite distribution and TV distribution is embedded in this network as just one specific kind of multimedia service.

A special variant of distribution services is of particular importance, and this is the realization of a limited form of control of the presentation by the user. This principle is known under the name 'Near Video on Demand' but it can be applied to any kind of multimedia distribution. The idea is to broadcast a large number of copies of the same information in parallel but with a different state relative to the actual time. In the case of a movie, several copies are broadcasted which differ in their starting time. By selecting a specific copy for his individual needs, the user can achieve some level of interactivity.

## 1.1.3 Generic Network Services

In a universal Broadband communication network which supports many multimedia services, there will always be some generic functionality which applies to several services. An example is the support of groups of closely collaborating people by providing a virtual infrastructure which appears as a separate private network (Virtual Private Network). Such a service is helpful in all service classes which were mentioned above. It provides for example user directories, mobility support or some protection against misuse by unauthorized users.

Another example of generic network features which can be combined with an arbitrary service is flexible billing and rerouting as it is supported today in the Intelligent Network for telephone networks (e.g. 800 service numbers).

Generally speaking, a good architecture for the support of multimedia services should also include such generic network services and provide a way to combine them with arbitrary multimedia services.

## 1.1.4   Network Capabilities to Support Multimedia Services

As reflected in the term multimedia, novel services would require the existence of more than one information type simultaneously. From the network point of view this means that either the different types of information should be handled as a single flow or that different streams of information belonging to the same user session should be kept synchronized so that applications run properly between the communicating users. In the first case, video, audio and data are encoded by the terminal and submitted to the network as a single information flow, while in the second case, the network provides the mechanisms for identifying and handling correlated information flows. Assuming that each flow is conveyed by a separate connection, the existence of more than one flow for a single session leads to the concept of multi-connection calls, a requirement which holds in general for multicast communications independently of the way information is submitted to the network. In multicast sessions a party should be able to join or leave at any time which means that each individual connection within a session should be handled as a single entity.

Another requirement stemming from the nature of multimedia services is inter-activity and negotiation with the network and/or the receiving terminal which from the network side is regarded as the ability of the user to control and modify its session not only at the establishment and release phase but also throughout the lifetime of the session. One typical example is the Video on Demand service where the user is expected to browse among films and video servers.

The support of real time video services and the existence of multimedia video servers accessed by several users has given rise to a number of network issues associated with the transfer and handling of user plane information. The prime issue, of course, is bandwidth. Video and in general moving image applications are typical paradigms of Broadband services requiring transfer rates massively above those offered today. Another issue is the definition of the exact bandwidth portion required by a video connection. Video is inherently a Variable Bit Rate (VBR) service, as opposed to the Constant Bit Rate (CBR) telephony service, and its traffic profile presents multiple time scale statistics that have led to the development of novel models [6]. Therefore, the user will not be able to accurately characterize the video traffic parameters. It can be expected that network operators will select VBR transmission for video to exploit the statistical multiplexing gain of packet mode transmission like the Asynchronous Transfer Mode (ATM) and/or the Internet protocol, and to take advantage of the higher signal quality and lower coding delay pertained to VBR transmission. However, in this case the network operator will have difficulties in pursuing an admission control decision. Issues become more evolving as one considers that end-to-end delay has to be constant to allow for real time video display and that the transmission rate has to be consistent with the parameters negotiated by the user and the network. To resolve these issues a number of accessing policies are proposed to smooth video traffic and several routing methods are developed to balance the video load in the network (see for example [7, 8]).

It can be easily understood that multimedia services impact on the network in terms of the network intelligence required to understand and process complex user requests as well as of traffic handling mechanisms undertaking the task to deliver information in a quality that is acceptable by the user. While the latter is a very important issue, this book focuses on the prime requirement. In the rest of this chapter a number of potential network technologies is reviewed with an aim to highlight the capabilities and the current status of each alternative with respect to multimedia service provision.

## 1.1.5  Review of Network Technologies

Video conference is an advanced interactive conversational multimedia service with high complexity in terms of network requirements as it involves audio, video and data transmission and requires real time information delivery and end-to-end synchronization. Therefore, it is a good paradigm for assessing the superset of requirements a typical multimedia service will impose to a network.

The emerging network standards for video conference can be broadly distinguished to those referring to circuit switched Integrated Services Digital Network (ISDN) and to those applied to packet-based networks as Local Area Networks with Internet Protocol and Broadband ISDN (B-ISDN) with ATM.

### 1.1.5.1 Integrated Services Digital Network (ISDN)

The Integrated Services Digital Network (ISDN) supports the transmission of audiovisual information through circuit switching at bandwidths quantized to be multiples of 64 (B-channel), 384 (H0-channel), 1536 (H11 channel) and 1920 kbit/s (H12 channel). The respective recommendation H.320 of ITU includes also the H.261 standard for video compression which experiences a buffering delay of four times the frame period plus the display delay due to picture skipping [9].

The main characteristic of audiovisual service provision in ISDN is that information flows on point-to-point connections with a fixed guaranteed bandwidth. To accommodate the multicast requirement of videoconference, H.320 defines the Multipoint Control Unit (MCU) which plays the role of a central conference center that controls the conference after all users get connected to it. Therefore, the approach followed is centralized as all streams are forwarded to the MCU which undertakes the task of selecting and combining streams for delivery to the participating parties. The centralized approach is further justified when considering the capabilities of the signaling system employed in ISDN. The constant rate of the ISDN connections implies that the encoder should produce video streams in varying quality, as video scenes and video compression depend on the video content. Also both the sender and the receiver should be kept synchronized to the same network clock.

Due to technical and partly historical reasons the control of connections in ISDN is performed via a separate communication channel, so that flow of user information and signaling messages has to be distinguished. Thus we have the so called out-of-band signaling, in contrast to the in-band signaling of Open Systems Interconnection (OSI) where connection control messages share the same channel with those bearing data. In the Protocol Reference Model of ISDN this separation is reflected into a vertical separation into planes. Each plane hosts separate protocols or protocol stacks [10]. A plane is given the name of the type of information it is responsible to handle. The current terminology has a user plane (U-plane) for user information, a control plane (C-plane) for control information including signaling, and a management plane (M-plane) for management information. Signaling information is exchanged between the user and the network at the so called User Network Interface (UNI) and also between the network nodes at the Network Node Interface (NNI). This separation between UNI and NNI signaling reflects the inability of older plain telephony terminals to execute the complex protocols required for the control of connections at the network nodes. The corresponding signaling systems for the UNI and NNI are the Digital Subscriber Signaling System No. 1 (DSS1) for access signaling and the Signaling System No. 7 (SS7) for network signaling [11]. Both DSS1 and SS7 do not offer the flexibility required for the description of sophisticated multimedia calls with multiparty and multiconnection characteristics. Their approach is monolithic in the sense that a service request is always treated as a single end-to-end object allowing no modifications, that is, a call contains one service, one connection and one channel which remain unchanged for the call lifetime. The ISDN signaling protocols are further discussed in Chapter 1.3.

## 1.1.5.2 Internet

The Internet Protocol was originally developed to support a simple connnectionless unreliable data transfer service over Local Area Networks (LANs) interconnected with routers. Quality of Service (QoS) was preserved by higher level protocols as TCP/IP which provides a packet retransmission mechanism suited well only to delay insensitive data. The increasing demand for real time multimedia services and the dramatic growth of Internet has led to the concept of an integrated services Internet which should be able to accommodate apart from the guaranteed delivery of packets, other more strict QoS requirements like delay, and end-to-end synchronization. To do so, the network should recognize and manipulate different service classes while the user should be able to request a service in a particular quality class. As there is no signaling for user-network communication in Internet, a control mechanism is currently developed for this purpose. The so called Resource ReSerVation Protocol (RSVP) provides the means to the user application for requesting a specific service class and for transferring this request to the Internet routers which reserve network resources for the corresponding flow accordingly [12]. The RSVP currently allows the user to select among three service classes each one offering distinct QoS characteristics: The 'guaranteed' service provides a bounded delay and no loss transmission for conforming traffic, the 'controlled load' service tolerates few losses by guaranteeing a minimum bandwidth and is oriented to delay adaptive applications, whereas the best effort service provides no guarantees and is suitable for delay and bandwidth adaptive services.

While in principle multicast communication constitutes a potential feature of any packet based network, the original Internet Protocol was designed to support only point-to-point interconnection of segments with internal multicast capabilities. As new services require the participation of more than two users in a session significant research effort over the last few years is spent on the definition of a multicast internet service model. Several multicast routing protocols are studied with the objective to find an efficient solution for determining the path of a message in terms of low cost routes, easily adaptive to changes in group membership, uniform traffic distribution to the network and minimization of the routing information kept in the nodes [13]. The multicast backbone (MBONE) started as a virtual network overlaid to Internet supporting one-to-many delivery of real time multimedia application through point-to-point encapsulated tunnels. It rapidly evolved to an experimental platform for the development of Internet integrated protocols with advanced capabilities not bounded to multicast but also others as for example the support of real time applications [14].

Attempts are currently made for integrating RSVP and Internet Protocol multicast with the signaling and connection management systems used in Broadband networks so that Internet Protocol applications can take full advantage of the ability of the latter to establish point-to-multipoint connections with quality of service guarantees [15]. This new area referred to as Internet Protocol over ATM concentrates the interest of the information technology community as it can successfully merge the two different concepts reflected in the computer communications approach adopted in Internet and the telecommunications approach used in ATM-based B-ISDN. The main working assumptions for this combination is that the network will mainly consist of fast ATM switches while user applications will be Internet-based. Selectively within the network and always at the access, Internet Protocol over ATM implementations will reside to provide interworking and router-like functionality.

## *1.1.5.3 Broadband ISDN*

Before ISDN had time to acquire much spread acceptance the newest technological developments have opened the way towards large scale Broadband networks. For Broadband ISDN (B-ISDN) the Asynchronous Transfer Mode (ATM) has been agreed as the adopted method of realizing the Broadband, integrated and of course digital aspects entailed in the B-ISDN acronym. As a future network, ATM based B-ISDN should strive not only for a common access through the so called Broadband UNI, but also for a really integrated networking infrastructure. In other words all users and services should exchange information through the use of ATM connections transporting ATM cells.

One of the main advantages of the ATM technology is the ability to support permanent and switched connections simultaneously. This has given rise to the concept of Virtual Paths (VPs) and Virtual Channels (VCs). The VP concept allows for applications of long time scale interconnections among groups of users of well known characteristics and statistic behavior. A typical application of this kind is the LAN interconnection governed by a single administration authority. The semi-permanent VP connections are subject to network management operations with appropriate medium or long time scale manipulations. Depending on the specific application supported VP connections may last for hours or even days and in this respect the VP network can be seen as an evolution of the SDH (Synchronous Digital Hierarchy) cross connect networks currently in use. The desirable advantages of the VP cross connect over their SDH and plesiochronous counterparts are given by the underlying ATM features: flexible and dynamic bandwidth allocation and manipulation as well as direct multiplexing/de-multiplexing procedures.

For switched ATM networks signaling occurs whenever VC connections have to be established, maintained, modified and released under user control through terminal and C-plane functions. Within the B-ISDN Protocol Reference Model the U- and C-planes are distinguished higher up than in the ATM layer [16]. Thus ATM cells are used invariably for the transport of user or signaling information and no direct way for identifying signaling traffic is provided by the ATM layer. To transfer signaling messages VP and/or VC connections have to be used between terminals and specific network nodes where call and connection related information can be processed. The architecture of the B-ISDN signaling protocols should be modular enough to allow the representation of complex advanced multimedia services while at the same time a high performance in terms of call set up delay should be guaranteed [17]. Up to now, widely used Broadband signaling protocols are monolithic in the sense that there are strict restrictions in the manner a service is structured. For example multi-connection and multiparty capabilities are not built in features in the Q.2931 signaling protocol standardized for the UNI (see Section 1.3.4). The current practice followed to accommodate the control requirements of novel multimedia services is based on what can be called 'soft' solutions: signaling protocols are enriched with additional messages and procedures that merely cover the requirements of the new services (see for example in Section 1.3.4 the *add party* procedure and related messages introduced in Q.2971 for the establishment of multiparty calls) or the network intelligence required to co-ordinate several simple calls and to put them under a common session umbrella is provided by dedicated servers in the network. Broadband signaling is further elaborated in Chapter 1.3.

Unlike N-ISDN, ATM multiplexes traffic from different sources in the form of fixed size packets known as cells. In principle the range of rates supported by B-ISDN is unlimited as a service can occupy as many cells as required for the transfer of its information. This form of statistical multiplexing enables a high utilization of network resources while resource reservation based on fixed rates is not excluded. In fact, the standardized ISDN rates are also supported in B-ISDN and provision has been taken for the interoperation of H.320 with H.321 its Broadband counterpart. The interoperability between narrowband and Broadband terminals is achieved by introducing an ATM Adaptation Layer Protocol 1 (AAL1) on top of ATM which provides a fixed rate service to the user equivalent to the narrowband B, H0, and H11/H12 channel.

Current developments based on ATM technology take account of the MPEG (Moving Pictures Expert Group) standards for coding audiovisual information. MPEG-1 achieves a rate of 1.5 Mbit/s in which a single stream of combined audio and video information with a common time base is transferred. MPEG-2 has offered the possibility to multiplex streams of independent time bases and targets at rates approaching the 10 Mbit/s. Further enhancements of MPEG, like MPEG-4 and MPEG-7, consider a scene as a composition of autonomous objects, making it possible to address more advanced requirements related to the handling of video information like interactivity and content based searching, and coding [18].

## 1.1.6   Intelligent Network and Alternative Approaches

Intelligence enables a system to exploit its basic capabilities to enrich and enlarge the set of offered services. The telephone network offers a mere switching facility and it is not suited for introducing new routing, charging and number translation capabilities in a stand alone fashion. At the end of the 60s the availability of program controlled switches opened the way to the introduction of switched based services, called supplementary services, emphasizing at the same time the potentiality and the limits of this kind of solution. Supplementary service provision requires the modification and the enrichment of the basic call process, performed by switches to provide the basic call service, in all the nodes supporting the new functionality. The high cost for the provision of new supplementary services to a large number of subscribers within a multi-vendor environment (a large number of switches needs to be modified) demonstrates the insufficiency of this initial solution.

As the provision of new network services is a very important topic for public network operators who are interested in increasing the revenues resulting from the exploitation of the already deployed network infrastructure, a lot of efforts have been made first to define proprietary solutions for properly adding intelligence to existing networks. This is exactly the point in time where the Intelligent Network concept appears. Traditionally the IN is the technology allowing the provision of new complex services on top of a Public Switched Telephone Network (PSTN) without affecting seriously the switching systems. The fundamental idea behind IN is to place the intelligence required for the provision of a complex service in dedicated IN servers instead of modifying the call processing software in every switch of the network. Doing so, the switch functionality is restricted to basic call processing and additionally to the identification of IN service calls and to the routing of these calls to the IN server.

The first IN solution was implemented in the United States at the beginning of the 80s to support the green number service. The green number service, also called 800 service, allows a called party reachable via an 800 number to pay for the cost of the incoming calls. This service is used, for example, by large companies to offer a free information service to their customers. The green number service has opened the way to a large number of services like the reverse charging, the call forwarding, the call deviation and many others.

All these services have the common characteristic of being accessible via a predefined numbering prefix. All public network operators have assigned to each IN service a specific numbering prefix that enables both the switching systems to distinguish between IN and normal calls and the intelligent service systems to discriminate among different services. The utilization of such service access codes depends, on one hand, on the requirement of making the IN services accessible to telephonic users and, on the other hand, on the need of integrating IN services with the existing switching services.

From the service point of view, the IN network can be seen as an overlay network. In fact, IN assumes an architectural separation between service control functions and service switching functions aiming at supporting network and vendor independence. To be more precise, the requirement of IN is not only the independence from the underlying network technology but also independence from the particular type of service being offered. This is one of the major points for public network operators willing to introduce new services minimizing the impacts on the already deployed network. As a large number of services can be provided combining low level service components, a good answer to this problem is the standardization of such service independent building blocks. Authentication, number translation, charging, routing are examples of service independent building blocks. Moreover, the utilization of a service independent call model and protocol between the switching and service nodes approaches the goal of a service independent architecture.

The above considerations justify the world-wide exploitation of IN during the last few years making it one of the most important developments in recent telecoms history [19, 20]. The survival of the IN concept into the next century largely depends on its capability to support complex multimedia services in a cost effective manner, and although IN standards are well specified for narrowband ISDN services, the situation is quite unstable for Broadband. The truth is that in the Broadband context the development of multimedia IN-based services relies on the realization of a high speed network with advanced signaling capabilities, a fact which induces changes in existing narrowband IN service creation and control systems. Therefore to preserve the fundamental IN principles of flexibility in design, deployment and customization of services, the logic inside the Service Control Point (SCP), that is the IN entity in which the processing of IN calls takes place, should be enhanced to provide a modular and efficient platform able to accommodate the complex processing requirements of new services. Furthermore, as the interactions between the network and the IN service system will become more intense, and the information exchanged more complex, it becomes evident that also the network related Service Switching Point (SSP) software should be upgraded.

Even with the above, the advantages of IN still prevail as the enhancements are performed towards a more efficient utilization of an upgraded underlying signaling system and do not in any way force a drastic and continuous modification of the deployed network infrastructure. Having this in mind several operators have chosen IN as a future safe solution that guarantees the robustness of their systems and the safety of their investments. In the following chapter the concept of the Intelligent Broadband Network is introduced and the motivation behind this choice for supporting advanced multimedia services is explained.

## 1.1.7   Intelligent Broadband Networks

This book gives a detailed technical description of the *Intelligent Broadband Network* approach which integrates Broadband ISDN and Intelligent Network concepts.

There are several basic assumptions on which this approach is based. The first and prime assumption is that future networks will be dominated by multimedia applications but will also have to work together smoothly with existing telecommunication networks and the Internet. The Intelligent Broadband Network provides an adequate infrastructure for high-quality multimedia services by using an ATM network as its basic communication infrastructure. The Broadband ISDN framework ensures interoperability with other telecommunication networks like the narrowband ISDN.

The second assumption is that there will be a split of functionality between the network and the end systems which requires significant enhancements compared to the current state of B-ISDN standards. The economic reason for this assumption is that the network operators will try to compete increasingly on the area of value-adding services rather than on the provision of plain transmission capacity, and therefore will be interested in network-specific support for advanced services. The technical reason is that some functionality like integrated and flexible billing for an advanced service is easier to achieve by an adequate network architecture than by network-independent applications.

The third and final assumption is that adequate network support for multimedia services requires a more abstract software architecture than the current Intelligent Network standards. Therefore, the Intelligent Broadband Network is much more than a simple transfer of Intelligent Network concepts to ATM networks. Instead, a new level of abstraction is introduced which is formulated in terms of an abstract object-oriented model. This object-oriented view of network resources and services is thoroughly integrated with the classical Intelligent Network architecture such that the new approach still supports all of the known IN concepts and therefore can be seen as a true extension of the IN architecture.

The material which is presented in this book can be read in different ways depending on the interest of the reader. Part 1 contains, besides this introduction, two other chapters which provide background material in the form of a summary of the state of the art in Intelligent Networks (Chapter 1.2) and Broadband

Signaling. This material can be skipped if the reader is already familiar with these concepts.

Part 2 describes the main technical approach to Intelligent Broadband Networks. After a general introduction into the used framework (Chapter 2.1), Chapter 2.2 defines the central functional models of the new architecture. Based on this architecture, Chapter 2.3 presents a version of the IN protocols which is adequate for Broadband multimedia services. Chapter 2.4 applies the architecture and the protocol to a number of example services. Chapters 2.2, 2.3 and 2.4 can be seen as the core architectural description, and should therefore be interesting to all readers. However, some of the technical detail in the protocol definition may be skipped on a first reading. Chapters 2.5, 2.6 and 2.7 discuss in detail a prototypical implementation of network elements for an Intelligent Broadband Network. These chapters are written for readers who want to understand the practical implementation issues but they are not necessary for understanding the remaining material. Chapter 2.8 summarizes the practical experience which was achieved in a big Europe-wide field trial and therefore seems to be attractive for all readers.

Part 3 concentrates on the interrelationships of an Intelligent Broadband Network with other network concepts. This material is important for everybody who wants to judge the strategic aspects of this branch of network evolution. Separate chapters are devoted to an overview of the interrelationships (Chapter 3.1), to TINA (Chapter 3.2), to narrowband networks (Chapter 3.3), to pure Broadband ISDN (Chapter 3.4) and to the Internet (Chapter 3.5).

Part 4 is concerned with the performance-driven design of Intelligent Broadband Networks. This material complements the architectural definitions with a theoretical analysis of performance aspects. Readers with a specific interest in this issue are recommended to read at least Chapters 2.1, 2.2, 2.3 and 2.4 first in order to get the appropriate background. Separate chapters in part 4 deal with an overview (Chapter 4.1), an introduction of the used methods and models (Chapter 4.2), an analysis of alternative architectural solutions (Chapter 4.3), a discussion of congestion and overload situations (Chapter 4.4), and a general analysis of the scalability of Intelligent Broadband Networks towards large networks.

At the end of the book, a number of useful appendices are put together which summarize the relevant protocol standards, the used notation for object-oriented models, details for the used queuing model as well as tables of system parameter values which were used in the performance evaluation.

# References

[1] Minoli D, Keinath R, *Distributed Multimedia through Broadband Communications*, Artech House, 1994

[2] ITU-T Recommendation I.374, *Framework Recommendation on Network Capabilities to Support Multimedia Service*, 1993

[3] ITU-T-Recommendation F.700, *Framework Recommendation for audiovisual/ multimedia services*, 1996

[4] ITU-T Recommendation I.211, *B-ISDN Service Aspects*, 1993

[5]   Ahamed S V, Lawrence V B, *Intelligent Broadband Multimedia Networks*, Kluwer, 1997

[6]   Conti M, Gregori E, Larsson A, *Study of the Impact of MPEG-1 Correlations on Video-Sources Statistical Multiplexing*, IEEE J. Select. Areas Commun, **14**, 1455-1471, 1996

[7]   Liew S, Tse C, *Video Aggregation: Adapting Video Traffic for Transport over Broadband Networks by Integrating Data Compression and Statistical Multiplexing*, IEEE J. Select. Areas Commun, **14**, 1123-1137, 1996

[8]   Maxemchuk N, Video Distribution on Multicast Networks, IEEE J. Select. Areas Commun, **15**, 357-372, 1997

[9]   Okubo S, Dunstan S, Morrison G, Nillson M, Radha H, Skran D, Thom G, *ITU-T Standardization of Audiovisual Communication Systems in ATM and LAN Environments*, IEEE J. Select. Areas Commun, **15**, 965-982, 1997

[10]  ITU-T Recommendation I.320, *ISDN Protocol Reference Model*, 1988

[11]  Minoli D, Dobrowski G, *Prinicples of Signaling for Cell Relay and Frame Relay*, Artech House, 1995

[12]  White P, *RSVP and Integrated Services in the Internet: A Tutorial*, IEEE Commun. Mag., **35**, 100-106, May 1997

[13]  Diot C, Dabbous W, Crowcroft J, *Multipoint Communication: A Survey of Protocols Functions, and Mechanisms*, IEEE J. Select. Areas Commun, **15**, 277-290, 1997

[14]  Almeroth K, Ammar M, *Multicast Group Behavior in the Internet's Multicast Backbone (MBONE)*, IEEE Commun. Mag., **35**, 124-129, June 1997

[15]  Armitage G, *IP Multicasting over ATM Networks*, IEEE J. Select. Areas Commun, **15**, 445-457, 1997

[16]  ITU-T Recommendation I.321, *B-ISDN Protocol Reference Model and its application*, 1992

[17]  De Prycker M, Vecchi M (eds.), *Special Issue on Signaling Protocols and Services for Broadband ATM Networks*, International Journal of Communication Systems, **7**, 71-160, 1994

[18]  Chiariglione L, *MPEG and Multimedia Communications*, IEEE Trans. Circuits Syst. Video Technol., **7**, 5-18, 1997

[19]  Thorner J, *Intelligent Networks*, Artech House, 1994

[20]  Lesley U, *The Intelligent Network Opens Up for Business*, Telecommunications, 30(11), 1996

# Chapter 1.2

# INTELLIGENT NETWORKS

This chapter gives an overview of the Intelligent Network (IN) concepts that provide the basis for the IN architecture. The focus is placed on the IN conceptual model, on the identification of functional and physical entities and on the understanding of IN functional models. Finally the role and objective of the IN standardization activity are outlined.

## 1.2.1 Basic Principles

To fulfil the requirement of providing added value services on top of a switching network the Intelligent Network architecture has been devised. The IN architecture identifies the network elements, the functional entities, the models, the finite state machines and the protocols that have to be integrated in the network to support the provision of IN services. The basic idea of IN is to keep the service logic, required for the control of IN services, and the corresponding database outside the switching network in intelligent nodes. A large number of switching nodes co-operate with a few intelligent nodes to make IN services accessible to users, as shown in Figure 1.2.1. The advantage of this solution is straightforward: new services can be simply provided by modifying a few service nodes. In this context to solve the problem of interworking between the equipment of different vendors the interface and protocols between switches and intelligent nodes have been standardized. The concept of out band signaling supported by the common channel Signaling System number 7 (SS7) was adopted by the IN architecture as a basis for the interconnection of switching and service nodes.

One of the major standardization bodies that has contributed to the definition of a common standard for the offering of IN services of top of existing network technologies is ITU-T. Analogously to what has been done for the standardization of signaling, the standardization of IN foresees the delivery of successive sets of recommendation of the Q.1200 series in correspondence with the identified IN capability sets. Each Capability Set (CS) identifies a set of IN service and service features that has to be supported. Hence, each CS comprises system and service requirements. The recommendations associated with a CS specify the architecture and the protocols needed to support this capability. The standardization bodies have initiated a standardization path that goes through 3 CSs (CS-1, CS-2 and CS-3). The IN services and feature sets supported by each CS are described in Section 1.2.7.

---

*Intelligent Broadband Networks.* Edited by I. S. Venieris, H. Hussmann
© 1998 John Wiley & Sons Ltd.

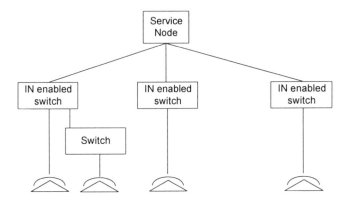

**Figure 1.2.1**
IN switching and service nodes

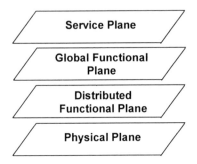

**Figure 1.2.2**
The planes of the IN Conceptual Model

The first result of this standardization activity has been the identification of a reference model suitable for the engineering process of IN, that is the IN Conceptual Model. The IN Conceptual Model allows to reduce the complexity of the IN service modeling, analyzing the problem from four different points of view, called planes. Each plane provides an abstraction of problem that can be studied independently from the other points of view. The four planes derive from the top down analysis of the IN architecture starting from the service point of view to the physical point of view. The four planes address the service aspects, global functionality, distributed functionality and physical aspects of an Intelligent Network [1] (see Figure 1.2.2).

## 1.2.2   Intelligent Network Conceptual Model Overview

According to [1] the IN Conceptual Model distinguishes the following four planes, each representing a different abstract view of the capabilities of an Intelligent Network:

- *Service Plane:* it describes services from the user point of view, without any reference to an IN based implementation. Each service is decomposed into basic Service Features (SFs) which are the base for the creation of other services;
- *Global Functional Plane*: it models the network from global perspective hiding the peculiarities related to the distribution of functional entities. This plane describes how the Service Independent Building Blocks (SIBs) can be composed in order to provide a specific service feature, identifying in this way a Global Service Logic (GSL). In addition, it describes the possible interactions between the Basic Call Process (BCP), which is in charge of launching and controlling IN services, and a particular SIB;
- *Distributed Functional Plane:* it models a distributed view of an IN structured network in terms of Functional Entities (FEs) which are able to realize various IN services. Each FE may perform a variety of Functional Entity Actions (FEAs) which, in turn, may be performed within different functional entities. However, a given FEA may not be distributed across FEs. The service components identified in the Global Functional plane are mapped onto sequences of particular FEAs performed in the FEs. Some of these FEAs result in Information Flows (IFs) between FEs;
- *Physical Plane:* it models the physical aspects of an IN structured network, including the detailed design of physical network elements, Physical Entities (PEs), and communication protocols.

Figure 1.2.3 depicts the result of the IN service decomposition through the four IN conceptual model planes. Examples of FEs in Figure 1.2.3 are SSF, SCF, SDF etc.; examples of PEs are SSP, SCP, SDP etc.

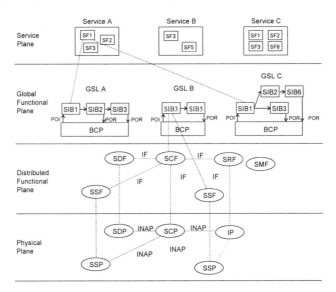

**Figure 1.2.3**
IN service decomposition

The Basic Call Process is a specialized SIB responsible for providing basic call capabilities like connecting and disconnecting calls or retrieving the call identifier. An IN service can be represented as a chain of SIBs connected to the BCP SIB. A Point Of Initiation (POI) is the BCP functionality needed to launch a chain of SIB. While a Point Of Return (POR) is the functionality needed to terminate a chain.

The Global Service Logic is the 'logic' that defines the order in which SIBs are chained together to accomplish a particular service. It is defined in terms of BCP interaction points (POI and POR), SIBs, logical connections, input/output data parameters, Service Support Data and Call Instance Data. It is the only element in the Global Functional plane that is specifically service dependent.

The definition of the Functional Entities and Physical Entities presented in Figure 1.2.3 is given in Section 1.2.4.

### 1.2.2.1 Relationships between Service plane and Global Functional planes

The abstract service descriptions of the Service plane are mapped onto service logic programs which can be executed by the Service Control Function (SCF, see Section 1.2.3). The process of transforming service descriptions into service logic is called Service Creation. The service features identified in the Service plane are realized in the Global Functional plane by a combination of Global Service Logic and SIBs including the Basic Call Process SIBs. This mapping is related to the service creation process.

### 1.2.2.2 Relationships between Global and Distributed Functional planes

The actual execution of service logic (by the Service Control Function, more precisely the Service Logic Execution Environment) is a distributed computation, involving several other functional entities (like the Service Switching Function (SSF), and Specialized Resource Functions (SRF)). Therefore the service logic as mentioned above is also called Distributed Service Logic. Each generic SIB is realized in the Distributed Functional plane by at least one FE or, in the general case, by the co-operation of several FEs.

### 1.2.2.3 Relationships between Distributed Functional and Physical planes

The FEs which are identified in the Distributed Functional plane determine the behavior of the Physical Entities (PEs) onto which they are mapped. Relationships (i.e. sets of information flows) between FEs, identified in the Distributed Functional plane, are specified as protocols in the Physical plane if they are exchanged between different physical entities. Distributed Service Logic (DSL) may be dynamically loaded into PEs and the mapping is related to the service management process.

### 1.2.2.4 Internetworking in the Distributed Functional plane

The Distributed Functional plane should be explicitly divided into several parts, each of which represents one functional network. Internetworking requires that relationships are defined between pairs of functional entities in different functional networks.

Each internetworking interaction between communicating pairs of functional entities is named an Information Flow (IF). The internetworking relationship between any pair of functional entities is the set of internetworking related information flows between them. The semantic meaning and information content of each IF needs to consider internetworking capabilities, network security and network integrity.

## 1.2.3   Intelligent Network Functional and Physical Entities

This Section introduces the functional and physical entities foreseen by the CS-2 which is the last standardized capability set. The physical entities of CS-2 are a superset of those defined in the CS-1.

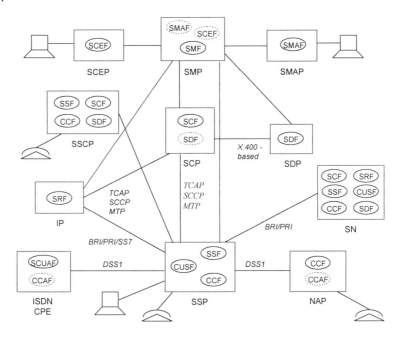

**Figure 1.2.4**
Example of a possible mapping of IN functional entities into IN physical entities supported by IN CS-2

The FEs defined in [2] related to IN service execution are:
- *Call Control Agent Function (CCAF)*: it provides access to users as it is the interface between user and network call control functions;
- *Call Control Function (CCF)*: it provides call/connection processing and control;
- *Call-Unrelated Service Function (CUSF)*: it supports the call unrelated

communication between users and the IN service logic. It is not present in CS-1;

- *Service Switching Function (SSF)*: it provides the set of functions required for the interaction between the CCF and the SCF;
- *Service Control Function (SCF)*: it controls call control functions in the processing of IN services;
- *Service Data Function (SDF)*: it contains customer and network data for real time access by the SCF;
- *Specialized Resource Function (SRF)*: it provides the specialized resources needed for the execution of IN services;
- *Service Control User Agent Function (SCUAF)*: it is in charge of user bearer unrelated interactions with the IN service logic. It is not present in CS-1.

The FEs related to IN service creation/management are:

- *Service Creation Environment Function (SCEF)*: it allows services provided by the IN to be defined, developed, tested and sent to the SMF. The output of this function involves service logic and service data templates;
- *Service Management Agent Function (SMAF)*: it provides an interface (e.g. screen presentation) between service administrators and the SMF;
- *Service Management Function (SMF)*: it allows deployment and provision of IN services and allows for the support of service monitoring and billing.

The physical entities defined in [3] for the support of IN CS-2 are the following:

- *Service Switching Point (SSP)*: it provides users with a network access (if the SSP is a local exchange), performs switching functionality and allows users to access IN services. It is able to detect user requests for IN services and to communicate with other IN physical entities. From the functional point of view, it contains the CCF. In addition, if the SSP supports call unrelated communication, the SSP contains also the CUSF;
- *Network Access Point (NAP)*: it includes only CCAF and CCF functional entities. It supports early and ubiquitous deployment of IN based services. It cannot communicate directly with a SCP;
- *Service Control Point (SCP)*: it contains the Service Logic Programs (SLPs) and data used to provide IN services. Multiple SCPs in the network may contain the same SLPs to improve the service reliability and share the IN traffic load. The SCP can access data in a SDP directly or through a signaling network;
- *Intelligent Peripheral (IP)*: it provides resources like voice announcements, voice recognition, dual tone multi-frequencies digit collection and is able to connect users to these resources. It supports flexible information interactions between a user and the network;
- *Service Node (SN)*: it can control IN services and interact with the user to exchange service related information. The SN is directly connected with one or more SSPs by means of a point-to-point signaling and transport connection;
- *Service Data Point (SDP)*: it contains the customer and network data which is retrieved during service execution;

- *Service Switching and Control Point (SSCP)*: it is the result of the combination of a SSP and SCP. The communication between service control functions and service switching functions within the SSCP is proprietary;
- *Enhanced ISDN Customer Premises Equipment (ISDN CPE)*: it provides the functions necessary for the operation of the access protocols by the user. It can contain the SCUAF for bearer unrelated interactions and CCAF for bearer related interactions.

Many different mappings of IN functional entities into IN physical entities are possible. Figure 1.2.4 shows an example of a typical but complex scenario. In fact, a very simple IN architecture could comprise only SSPs, containing the CCF and the SSF FEs, and a single SCP with an internal or external SDF. One or more IPs could be present to allow a poor interaction between the user and the network. If a more enhanced user service interaction is required, the SSPs could be enriched with a CUSF.

In the following chapters, some of the most relevant FEs will be discussed in more detail.

## 1.2.3.1 Call Control Function

The CCF is the function that provides call/service processing and control. According to [2], the CCF supports the following functionality:
- It establishes, manipulates and releases call/connection as requested by CCAF;
- It provides the capability to associate and relate CCAF functional entities that are involved in a particular call/connection instance (that may be due to SSF requests);
- It manages the relationship between CCAF functional entities involved in a call;
- It provides trigger mechanisms to access IN functionality (e.g., passes events to the SSF);
- It manages basic call resource data (e.g., call references).

## 1.2.3.2 Service Switching Function

According to the [2], the SSF provides the set of functions required for interaction between the CCF and a SCF:
- It extends the logic of the CCF to include recognition of service control triggers and to interact with the SCF;
- It manages signaling between the CCF and the SCF;
- modifies call/connection processing functions (in the CCF) as required to process requests for IN provided service usage under the control of the SCF;
- It supports the relay case, in which the SSF ensures the relay of information between the SCF and SRF possibly using the Out Channel Call Related Interaction (OCCRUI) capabilities.

## 1.2.3.3 Service Control Function

The prime function of the SCF is the execution of Service Logic provided in the form of service logic programs. It is this capability which enables the SCF to provide the mechanisms for introducing new services and service features independent of switching systems. In summary, the SCF:

- Interfaces and interacts with Service Switching Function (SSF)/Call Control Function (CCF)/Call Unrelated Service Function (CUSF), Specialized Resource Function (SRF) and Service Data Function (SDF) functional entities;
- Contains the logic and processing capability required to handle IN call related and IN call unrelated service requests;
- Provides security mechanisms for the purposes of internetworking;
- Interfaces and interacts with other SCFs for the purposes of distributed service control.

The above is a summary of the definition provided in [2].

## 1.2.3.4 Specialized Resource Function

The SRF provides the following functions:
- Interfaces and interacts with SCF and SSF/CCF;
- Contains the logic and processing capability to receive/send user information;
- Converts (if necessary) information received from users.

## 1.2.3.5 Service Data Function

The SDF contains user/terminal and network data for provision and operation of IN services. SDF functionalities are as follows:

- Storing, managing and accessing data functionalities;
- Data exchange functionalities;
- Security functionalities (e.g. authenticate users, assign user's access rights, block data access, etc.).

Additionally the SDF:

- Interfaces and interacts with other SDFs when required, making the actual data location in the network transparent;
- Provides security mechanisms, to enable secured information transfer across the boundary between networks;
- Provides data support for security services. This can be used by the SDF itself for secured data management;
- Facilitates the co-operation of a robust recovery mechanism for copying of data (e.g. in the case the SDF is unavailable);
- Provides data access scripts (methods) which may be invoked by the SCF in order to simplify the information transfer via the SCF-SDF interface.

## 1.2.3.6 Service Management Function

The SMF functions can be grouped into five categories: Service Deployment Functions, Service Provisioning Functions, Service Operation Control Functions, Billing and Service Monitoring Functions.

Some examples of these functions are given as follows:

- The Service Deployment Functions include service scripts allocation, service generic data allocation, specialized resource data introduction/allocation, service testing, etc.;
- The Service Provisioning Functions collect service subscriber specific data and administrate it in subscriber databases and contract databases. The function translates the service and subscriber data into network specific data;
- The Service Operation Control Functions include service maintenance, software maintenance, service generic data/customer specific data updating, signaling routing data updating, specialized resource data updating, security functions, etc.;
- The Billing Functions include collecting Charging Records, generating/storing Charging Records, modification of tariffs;
- The Service Monitoring Functions include initiating measurements, collecting, analysis and reporting measurement data and fault monitoring data.

## 1.2.4 CCF/CUSF/SSF Functional Models

IN CS-1 foresees that an SSP is composed by the CCF and the SSF functional entities. While IN CS-2 envisages the presence of an additional functional entity, the CUSF, to deal with call unrelated aspects. Figure 1.2.5 depicts the CCF/CUSF/SSF functional models in a 'single ended' scenario. (The definition of 'single ended' is given in 1.2.10.)

As defined in [2], the SSF comprises of four managers each specialized in the accomplishment of specific functionality:

- *SCF Access Manager (SCF AM)*: it regulates the interaction between external FEs with the local FE. In particular, it is able to dispatch outgoing messages towards the destination FE and to properly send them through the appropriate protocol stack;
- *IN Switching Manager (IN-SM)*: it interacts with the SCF via the SCF AM during the IN service provision. In particular, it is in charge of providing the SCF with a view of SSF/CCF call/connection processing activities and of enabling the SCF control on SSF/CCF capabilities and resources. It is able to recognize which call/connection processing events have to be reported to the SCF. It performs the IN Switching State Model (IN-SSM) processing;
- *Feature Interaction Manager/Call Manager (FIM/CM)*: it supports multiple concurrent instances of IN service logic instances and non-IN service logic instances on a single call. It co-operates with the BCM and the IN-SM to provide the SSF with a unified view of the processing associated with a single call;

- *Basic Call Manager (BCM)*: it is not a functional entity. It provides an abstraction of the basic call/connection processing required to establish a communication path between users. It detects basic call/connection control events that can lead to the invocation of an IN service logic instance. This is called Detection Point (DP) processing.

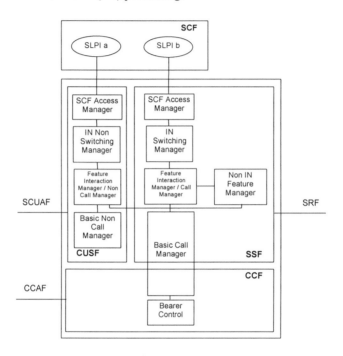

**Figure 1.2.5**
CCF/CUSF/SSF functional models

The activity of the BCM overcomes the SSF boundaries to realize some CCF related functionality such as the Basic Call State Model (BCSM) processing. The CCF is also in charge of signaling which is performed by the Bearer Control.

All the interfaces between the managers located within a physical entity are not subject to standardization as they are not externally visible.

As a general remark, it should be noticed that, within the CS-1, the role of IN-SM has not been emphasized since the small complexity of CS-1 IN services modeling has not stimulated its real exploitation.

The CUSF functional entity is present within SSPs whenever the non-call related user to service interaction is supported. In this case, new additional managers have to be introduced:

- *IN Non Switching Manager (IN-NSM)*: it interacts with the SCF via the SCF Access Manager during the IN service provision. In particular, it is in charge of providing the SCF with a view of CUSF call unrelated processing activities and of enabling the SCF control on CUSF capabilities and resources. It is able to recognize which call unrelated processing events have to be reported to the SCF;

- *Feature Interaction Manager/Non Call Manager (FIM/NCM)*: it supports multiple concurrent instances of IN and non-IN service logic instances on a single association. It co-operates with the BNCM and the IN-NSM to provide the SSF with a unified view of the call unrelated processing internal to the CUSF for an association;
- *Basic Non Call Manager (BNCM)*: it provides an abstraction of an association for call unrelated interaction between the user and the network. It detects basic call unrelated events that can lead to the invocation of an IN service logic instance and manages resources required to support basic call unrelated control. It is responsible for the Basic Call Unrelated State Model (BCUSM) processing.

## 1.2.4.1 CCF/CUSF Basic State Models

The IN CS-2 deals with basic state models in the CCF and CUSF, i.e. the BCSM and the BCUSM.

The BCSM provides a high level description of CCF activities related to the establishment, modification and releasing of communication paths between users. For this purpose, the state model identifies a set of basic CCF activities related to call and connection processing which the IN-SM and FIM/CM are interested in. The BCSM can be seen as a tool for representing CCF activities.

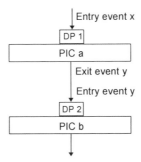

**Figure 1.2.6**
BCSM detection points and points in call

As shown in Figure 1.2.6, a BCSM is composed of:

- *Point in Calls (PICs)*: they identify CCF activities associated with one or more basic call/connection states of interest for the IN service logic;
- *Detection Points (DPs)*: they indicate states in the basic call/connection processing where the control can be transferred to the IN service logic with or without the CCF processing suspension. If the CCF processing is suspended, the DP is called a Request DP (DP-R) otherwise it is a Notification DP (DP-N). A DP can be armed, i.e. the monitoring of a CCF processing state can be requested, statically (on a configuration base) or dynamically (on IN

service logic request). In the first case the DP is referred as a Trigger DP (TDP) in the second case as an Event DP (EDP);

- *Transitions*: they indicate the normal flow of basic call/connection processing from one PIC to another;
- *Events*: they can be classified as entry events and exit events depending if they cause the BCSM transitions into PICs or they represent the result of PIC processing.

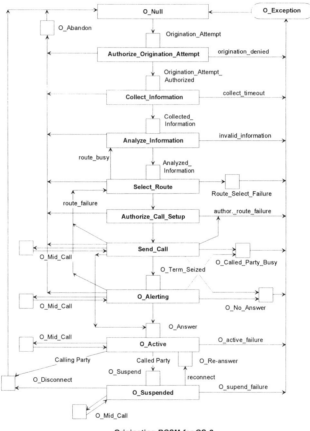

Originating BCSM for CS-2

**Figure 1.2.7**
Simplified version of CS-2 O_BCSM.
Reproduced with permission of ITU from recommendation Q.1224 full source of original publication.

Both the BCSMs defined in the CS-1 and CS-2 reflect the separation between originating and terminating portions of calls. Consequently the CCF processing of an IN 'half call' can be modeled by means of its originating or terminating half BCSM (O_BCSM and T_BCSM), depending on which CCF processing event triggered the IN service logic execution. The originating half of the BCSM models the calling party signaling activity, while, the terminating side models the called party signaling activity. In any case the two BCSM are not independent but strictly correlated as the events processed by one BCSM can be reflected in the corresponding events in the other BCSM. Each BCSM instance is handled by an associated BCM instance. Figure 1.2.7 illustrates a simplified version of the CS-2 originating half of the BCSM, the complete BCSM model is described in [2].

The BCUSM deals with the call unrelated associated connection oriented interaction between the user and the IN service logic based on ISDN signaling procedures. Analogously to the BCSM, the BCUSM provides a high level description of CUSF activities required to establish and maintain an association between users and service processing. The modeling of BCUSM is very similar to the BCSM one. The components used to describe a BCUSM are DPs and Point In Association (PIA), the BCUSM equivalent of PICs. Transactions and events are not required as they are evident. Figure 1.2.8 depicts the CS-2 BCUSM proposed in [2].

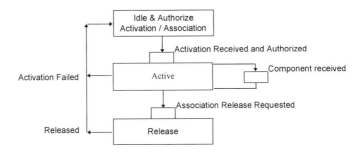

**Figure 1.2.8**
CS-2 BCUSM

## 1.2.4.2 SSF Switching State Model

According to [2], the IN Switching State Model (IN-SSM) provides an object-oriented description of SSF/CCF IN call/connection processing in terms of IN call/connection states. It provides a framework for describing the scope of view and control of SSF/CCF activities offered to an SCF. The extent to which the IN-SSM is visible to the SCF is defined by the information flows identified between the SSF/CCF and the SCF.

As introduced in Section 1.2.4, in CS-1 the role of the IN-SM has not been emphasized and consequently neither that of IN-SSM, though the technical base for their future exploitation was given. In IN CS-2, a more powerful IN-SSM has been defined which focuses on connection control issues. It therefore contains objects that are abstractions of switching and transmission resources.

The CS-2 IN-SSM uses the Call Configuration (CC) model as a tool to represent the CCF activities. (Note that while CC denotes the Call Configuration in the CS-2 IN-SSM, everywhere else in this book CC stands for Call Control with the exception of Chapter 3.2 where it stands for Connection Coordinator in the context of TINA). The CC is a model used to categorize the status of one or more network connections. It is based on the Connection View (CV), in the sense that CV objects provide the SCF with a generic view of call processing and each Call Configuration models a CV state in an SSF. This IN-SSM model has been developed for dealing with narrowband connections.

The purpose of defining CCs is to produce a set of examples used to describe IN call manipulation services. Figure 1.2.9 illustrates the notation and the objects used in describing the CCs. A CC shows the connectivity between a local leg and one or more remote legs.

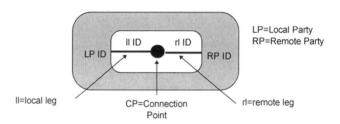

**Figure 1.2.9**
Representation used in a Call Configuration

Figure 1.2.10 illustrates the Call Association Object: it is the representation of multiple calls associated with a particular user as perceived by the Serving Node associated with that user. The Call Association Object allows the graphical representation of the condition when a user is engaged in more than one call.

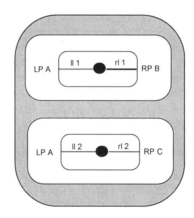

**Figure 1.2.10**
Representation of a Call Association Object

This approach has been developed to provide IN multiparty handling capability, and it can be used as a tool for associating different calls.

## 1.2.5   SCF Functional Model

As defined in [2] the SCF provides a service logic execution environment (SLEE) on which a service logic processing program (SLP) runs. When the SCF interacts only with entities which are part of an IN-structured network, the SCF model is that shown in Figure 1.2.11. It comprises the following entities:

- *Service Logic Program Library (SLPL):* this is a library of service application programs, which are used to realize service processing. SLPs contain logical constructs for controlling the flow of service execution and statements for invoking functional routines, in the SCF, which facilitate access to network resources and data necessary for service execution;

- *Service Logic Execution Environment (SLEE)*: it is the component of the SCF architecture where the SLPs are executed. The SLEE is responsible for managing simultaneous invocation and execution of multiple SLPs. It is composed of:

  —   *Service Logic Execution Manager (SLEM)*: It handles the total service logic execution. It contains the Service Logic Selection/Interaction Manager (SLSIM), Service Logic Program Instances (SLPIs) and the Resource Manager and, in order to support SLPI execution, interacts with the SCF Data Access Manager and the Functional Entity Access Manager (FEAM). It selects and executes the appropriate SLPIs and maintains transient data associated with them, executes functional routines in support of SLPI execution, manages SLPI access to the SDF via the SDF Access Manager and manages the exchange of information between SLPIs and components of other FEs via the FEAM. A SLPI is a dynamic entity that controls the flow of service execution and invokes SCF functional routines. The Resource Manager takes care of reserving and releasing all resources requested by specific SLPIs.

  —   *SCF Data Access Manager*: this entity provides the functionality needed for the storage, management and access of shared and persistent (i.e. persisting beyond the lifetime of a SLPI) information in the SCF. This entity is also responsible for accessing remote information in SDFs. The SCF data are contained in the Service Data Object Directory (SDOD) and in the IN Network-wide Resource Data (IN-NRD). The SDOD helps locate service data objects in the network in a manner transparent to the SLEM and its SLPIs. The IN-NRD contains information about the location and capabilities of resources in the network, that are accessible to the SLPIs.

  —   *Functional Routine Manager (FRM)*: it interacts with the Service Management Function (SMF) and is responsible for receiving the various functional routines from the SMF and distributing them to the Functional Routine Library. It is also responsible for adding, deleting or suspending a particular functional routine, under the control of the SMF.

  —   *Functional Routine Library (FRL)*: it is a library of functional routines which, in turn, provide functionality that causes a sequence ofFEAs to be performed in support of service execution. This sequence of FEAs is the functionality defined for a SIB in the Global Functional Plane.

— *Functional Entity Access Manager (FEAM)*: it provides the SLEM and all other FEs with a message handling functionality. In other words, it is the interface between the SLEM and all other FEs, in a manner transparent to the SLPIs and compliant to OSI structures and principles, at the same time providing reliable message transfer, sequential message delivery, capability of correlation of message request/response pairs, capability of association of multiple messages with each other.

— *SLP Manager*: this entity interacts with the Service Management Function (SMF) and is responsible for receiving and distributing the various SLPs from the other FEs. It is also responsible for adding, deleting or suspending a particular SLP, under the control of the SMF.

— *Security Manager*: this entity provides secure access to the functions of the SCF. It generates/verifies security parameters for outgoing/incoming messages in order to provide data authentication, data integrity and confidentiality for a specific interface.

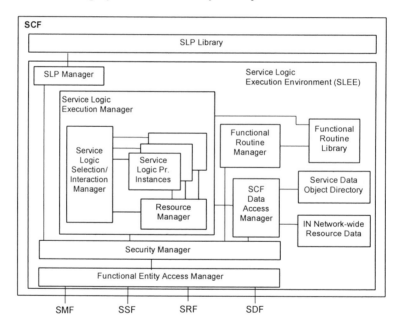

**Figure 1.2.11**
SCF functional model

## 1.2.6   SRF Functional Model

The Specialized Resource Function (SRF) provides specialized resources that may be needed for the execution of IN services [2]. The SRF interfaces with the

SCF and the SSF (and with the CCF). It may contain logic and processing capability to receive information from users and convert it to a usable (digital) form or to present information to users. It may also contain functionality similar to the CCF to manage bearer connections to the specialized resources.

Some examples of resources that may be managed by the SRF follow:

- Dual tone multi-frequency (DTMF) receiver;
- Tone generator/announcements;
- Message sender/receiver;
- Synthesized voice/speech recognition devices with interactive prompting facilities;
- Text to speech synthesis;
- Protocol converters;
- Audio conference bridges;
- Information distribution bridges.

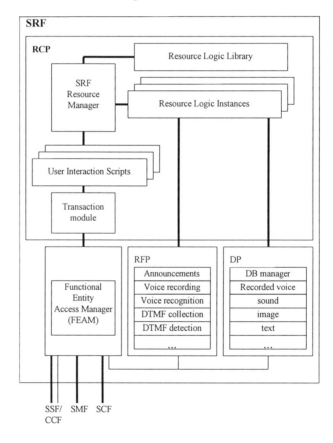

**Figure 1.2.12**
SRF functional model
Reproduced with permission of ITU from recommendation Q.1224 full source of original publication.

The SRF provides a set of special resources (which often represent special hardware devices). For service processing the SRF has logical relationships with the SSF/CCF and the SCF. The SCF controls the connection between the SSF/CCF and the SRF. SRF actions are activated by request from the SCF - the SRF never starts an action by itself. The SRF is managed by the SMF to place resources in service or out of service, to update the database of announcements, voice, sound etc.

Figure 1.2.12 shows the components the SRF consist of:

- *Resource Control Part (RCP)*: this part of the SRF contains the service logic and controls the service procedure using the other blocks. When the RCP receives a request for a specialized resource it invokes the internal resource controller which manages pairs of resource from the Resource Function Part (RFP) and data from the Data Part (DP) and decides on whether the resource request should be accepted or rejected. (Note that while DP denotes the Data Part in the context of SRF, everywhere else in this book DP stands for Detection Point). There is a resource controller for each resource type. The RCP consists of the following entities:
  - — *Resource Manager*: it provides the necessary functionality to search for resources, allocate and de-allocate them, manage the status of the resources and control their action.
  - — *Transaction Module:* this entity has capabilities to detect transactions from the communication links and to route the transaction to the corresponding application scripts.
  - — *User Interaction Scripts:* they provide the SCF with a view of the different resources and their possible actions. The scripts allow the grouping of user interaction parts of a service into blocks in an efficient way. The transition from one part to another is triggered by either internal events (error conditions, time-outs, etc.) or external decisions (user choice, database results). Each script represents a generic template, which may be dependent on parameters.
  - — *Resource Logic Library*: it provides the logic for the different user interaction scripts.
  - — *Resource Logic Instances:* they are instances of the SRF resources that are necessary for the execution of resource requests.
- *Resource Function Part (RFP)*: it collects together resources and functional elements of resources;
- *Data Part (DP)*: it consists of a Database Manager and a database that contains various data that is necessary for the execution of the specialized resources like: voice, sound, images, text, etc.;
- *Functional Entity Access Manager (FEAM)*: this component of the SRF provides the RCP with message handling functionality. It is the interface between the RCP and the other FE in the IN network in a transparent and OSI-compliant manner. It provides the SRF with: reliable message transfer, sequential message delivery, request/response correlation and the capability to associate events from the different FE with each other. The FEAM

communicates with the SSF/CCF via the standard signaling protocol and with the SCF via the INAP application protocol.

Interaction between the SCF and the SRF is either call-related or call-unrelated. The call-related interaction must be preceded by the establishment of a relationship between the SSF and the SCF and (usually) the set-up of one or more bearer connections between the SRF and the user. The user interaction consists of a chain of 'Questions and Answers', that may, for CS-2, be bundled in User Interaction Scripts. These user interaction scripts are executed as an indivisible block under control of the SRF. During the execution of a script the SRF may request additional information from the SCF. The result of the script is reported back to the SCF. While, the call-unrelated interaction is possible without a SCF/SSF relationship, it is used mainly for resource monitoring and administration by the SCF.

## 1.2.7   SDF Functional Model

The SDF comprises the components depicted in Figure 1.2.13.

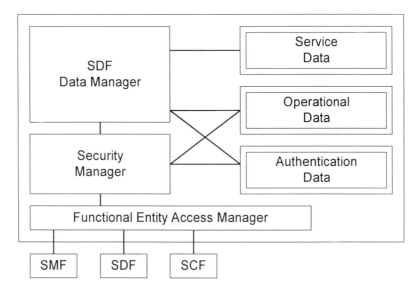

**Figure 1.2.13**
SDF functional model
Reproduced with permission of ITU from recommendation Q.1224 full source of original publication.

The SDF components are described below, as defined in [2]:
* *SDF Data Manager*: it provides the functionality needed for storing, managing and accessing information in the SDF;

- *Security Manager*: it provides secure access to the different types of data held in the SDF, for example, denied access to the data for unauthenticated users. The security manager also provides the functionality needed for providing secure access to the functions of the SDF by other Functional Entities (SCF, SDF);
- *Functional Entity Access Manager*: it provides the functionality needed by the SDF data manager to exchange information with other functional entities, i.e., SCF, SDF and SMF, via messages. This message handling functionality should provide reliable message transfer, ensure sequential message delivery, allow message request/response pairs to be correlated and allow multiple messages to be associated with each other;
- *Data Types* handled by SDF are:
  — *Service Data* used for the provision of a service, e.g., a subscriber profile, service provider agreements;
  — *Operational Data* used by the SDF itself for operational and administrative purposes, e.g., references to an object class, access control data;
  — *Authentication Data* used to authenticate a user that accesses the database through a SCF, e.g., a PIN code, the value of a counter for failed authentication.

## 1.2.8   SMF Functional Model

The SMF is shown in Figure 1.2.14.

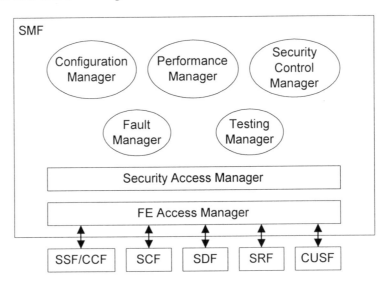

**Figure 1.2.14**
SMF functional model.
Reproduced with permission of ITU from recommendation Q.1224 full source of original publication.

As defined in [2], the SMF comprises a Configuration Manager, Performance Manager, Fault Manager, Security Control Manager, and Testing Manager.

- *Configuration Manager*: this is the component responsible for changing configurations for service deployment and service operation control;
- *Performance Manager*: it is the component responsible for gathering the appropriate performance information;
- *Security Control Manager*: it provides for logging of invalid access attempts and generation of alarms, activation/deactivation of appropriate security features to enable correct interface operation, and distribution of security related information;
- *Fault Manager*: it analyses and correlates alarms and performs appropriate tests to determine root cause of fault conditions;
- *Testing Manager*: it is responsible for the invocation and management of testing capabilities;
- *Security Access Manager*: it provides the functionality needed for secure access.

## 1.2.9   Intelligent Network Protocols

Existing Signaling System number 7 (SS7) lower layer protocols are used between the physical entities of Figure 1.2.4 to carry the application layer messages required by IN services. Consequently, only the application layer messages need to be standardized for each IN capability set.

The proposed underlying protocol platforms are illustrated in Figure 1.2.15. On the SSP/SCP interface the application message exchange takes advantage of the Transaction Capability Application Part (TCAP) protocol on top of the Signaling Connection Control Part (SCCP) and Message Transfer Part (MTP) protocols of SS7. A more detailed presentation of the involved signaling protocols appears in Chapter 1.3.

In a narrowband scenario, the SSP/IP interface is a Basic Rate Interface (BRI), Primary Rate Interface (PRI) or SS7 interface. The SSP can act as a relay node with respect to the SCP/IP communication. In this case, if a BRI/PRI protocol platform is used between the SSP and the IP, the application layer information related to the SCF/SRF communication are transported by means of FACILITY messages on the ISDN D channel set up between the SSP and the IP. In fact, the SCF/SRF application layer information are embedded in the FACILITY Information Element which can be transported by Q.931 messages (e.g. SETUP, CONNECT, RELEASE, FACILITY).

The application layer message exchange between the SSP and the SN physical entities is based on the utilization of common element procedures of Q.932 over an ISDN D channel of a BRI or PRI interface.

On the SCP/IP and SCP/SDP interface the application message exchange takes advantage of the same protocol stack utilized at the SSP/SCP interface.

The IN Application Protocol (INAP) defines the IN application layer messages exchanges between the SSF, SCF, SRF and SDF functional entities (see Figure 1.2.15). According to X.208, the INAP specification is realized with

ASN.1 notation. The encoding of the INAP messages is given by the Basic Encoding Rules (BER) as recommended in X.209 with the restrictions imposed by Q.773. In general, INAP utilizes some generic Application Service Elements (ASEs) such as the Transaction Capabilities (TCs) and the Remote Operations Service Elements (ROSEs). INAP defines an operation type in relation to each information flow exchanged between two communicating functional entities. The arguments of INAP operations are the information elements defined for the corresponding information flow. Within CS-1 and CS-2 the SSF/SCF information flows and consequently the INAP messages have a semantic that is mainly oriented to describe operations on the BCSM or BCUSM.

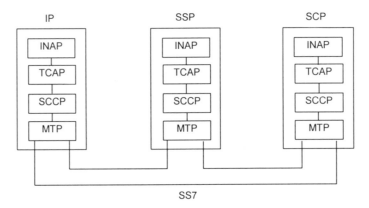

**Figure 1.2.15**
The IN protocol stacks

## 1.2.10  Intelligent Network Standardization

The international standardization of IN started in 1987 in ITU-T and ETSI with the first CS (CS-1). The work of ITU-T on CS-1 has been particularly difficult as different already deployed proprietary solutions were proposed for the standardization.

ITU-T provides a set of recommendations for each capability set that reflects the IN conceptual model points of view. Section A.1 of Appendix A shows the recommendations available for each capability set. Moreover, the Q.120y series provides the foundation for the definition of IN capability sets.

Basically the activity on IN has been subdivided in three capability sets. CS-1 comprises those IN services that can be enjoyed by users of a fixed telephone network like POTS and ISDN. CS-2 takes into account personal and partially terminal mobility services, IN service interworking, IN management and IN service creation. While, CS-3 mainly covers the full support for mobile and personal communication systems for users of both fixed and mobile networks as foreseen by the Universal Mobile Telecommunication System (UMTS) and by the International Mobile Telecommunication 2000 (IMT-2000), and integration with TMN and B-ISDN.

The CS-1 activity started in 1989. A first version of the specifications was approved in 1992 and a revised version was released in 1995. The activity on CS-2 started in 1992 and it has been completed in 1997. Finally, the activity on CS-3 started in 1996 and should end about the year 2000.

Services and service features of each IN CS fall into the Single Point Of Control or Multiple Point Of Control category and into the Single-Ended or Multi-Ended category, defined in [4]. The single-ended and single point of control services are Type A services, otherwise services are of Type B.

Single point of control refers to a control relationship where the same aspects of a call are influenced by one and only one SCF at any point in time. While multipoint of control is the ability for multiple service instances to interact with a single call.

Single-ended refers to the possibility for a service logic instance to directly influence only the processing of an isolated 'half call' (i.e. the originating or terminating BCSM) in the SSF/CCF. The other 'half call' can be only indirectly influenced by the information propagated by the controlled 'half call'.

In the following the characteristics of the IN capabilities sets are described.

## 1.2.10.1 CS-1

The CS-1 is the first result of the IN standardization process. The related recommendations are those of Q.121x series (see Table A.2 in Appendix A). The CS-1 aims at supporting only Type A services, as Type B services are considered too complex from the operation, the implementation and the control point of view. The target networks for the provision of CS-1 services are PSTN and ISDN.

As far as the telecommunication services is concerned, the CS-1 supports the following type of services [4]:

- Number translation services like freephone, call forwarding and premium rate;
- Flexible billing services like split charging and premium rate;
- Screening services;
- Incoming and outgoing calls restriction services.

Moreover, it includes the capabilities for the support of user interaction with an intelligent peripheral. This service feature is needed for the provision of many CS-1 services. The CS-1 also covers the basic service management capabilities related to the addition, modification or deletion of service specific data in the SSF, SCF, SDF and SRF on network operator or service customer request. The IN functionality required to support IN CS-1 services is as follows:

- End user access to call/service processing;
- Service invocation and service control;
- End user interaction with service control.

## 1.2.10.2 CS-2

The Intelligent Network CS-2 is the second standardized stage of IN. The recommendations of Q.122x series address all topics related with this capability set (see Table A.3 in Appendix A). The architectural concepts for the creation and provision of services, including telecommunication services, service management service and service creation services are defined. As in the CS-1, CS-2 services are single point of control services but they can involve more than two parties. IN CS-2 services are considered Type B services in the sense they can be invoked at any point during the call on behalf of one or more users. This means that during the call several parties may be added or dropped dynamically, requiring call configuration modification.

The CS-2 is a superset of CS-1. This means that all the IN capabilities supported by the CS-1 are also supported in CS-2. Moreover, the CS-2 provides some functionality that enable the IN architecture evolution towards the CS-3 and in particular towards IMT-2000 for the support of mobility.

As far as the CS-2 support for telecommunication services is concerned, new service features have been introduced [5]:

- Internetworking between IN services like the freephone service and the premium rate service;
- Personal mobility services like user authentication and user registration;
- Call party handling services such as call transfer and call waiting.

Some telecommunication services already introduced by CS-1 are extended in the CS-2. These services are: flexible routing, flexible charging, out channel call related user interaction, multi-party control (more than three parties can be involved in an IN service exploiting the Connection View State capability which allows the SCF to control multiple basic calls in the SSF/CCF), call unrelated user interaction, interaction between services, and service interworking over network boundaries (SCF to SCF and SDF to SDF relationships).

The CS-2 covers service management service aspects like service deployment, provisioning, customization and management during the service utilization. In addition, the service creation aspects are here addressed for the first time.

To achieve the IN CS-2 objectives new basic functionality has been identified. The most interesting functionality is described below:

- The user is able to invoke another service during the call active phase as the Mid-Call event is modeled;
- A very complex Call Party Handling functionality has been introduced to manage a multiparty call. The IN-SSM focuses on the connection control aspects as it uses the Call Configuration model. A call configuration shows the connectivity between a local leg and one or more remote legs;
- The Call Unrelated User Interaction (CUUI) is supported. This functionality is utilized by services like terminal registration and out channel based UPT location registration. The introduction of CUUI requires new functional entities (CUSF, SCUAF) and a new Basic Call Unrelated State Model;
- The Out Channel Call Related User Interaction (OCCRUI) allows the user to communicate with a service logic instance within the context of a call. The introduction of OCCRUI requires signaling interworking between basic call signaling and INAP and the introduction of new INAP IFs (Report UTSI and

Request STUI), where UTSI means User To Service Interaction and STUI Service To User Interaction, and new IEs.

### 1.2.10.3 CS-3

The work on the CS-3 is based on a phased approach. Two steps have been defined. The CS-3 Step 1 shall address the following topics:

- Evolution of CS-2;
- Multiple point of control;
- Service interaction;
- Number portability;
- IN/ISDN interworking;
- Support for narrowband mobility services: Universal Personal Telecommunication (UPT), GSM, terminal mobility, Cordless Terminal Mobility (CTM);
- Support for Broadband services involving only point-to-point monoconnection calls and point-to-multipoint calls where the root interacts with IN.

While Step 2 will address:

- Broadband mobility services, IMT-2000;
- Complete integration with B-ISDN;
- IN/Internet interworking;
- Internetworking procedures extension;
- Integration with TMN.

The activity on CS-3 is an ongoing activity. The CS-3.1 standardization activity should be completed at the end of 1998, while the CS-3.2 related activity should be finalized for the year 2000.

## References

[1] ITU-T Recommendation Q.1201, *Principles of intelligent network architecture*, 1992

[2] ITU-T Recommendation Q.1224, *Distributed functional plane for intelligent network CS-2*, 1997

[3] ITU-T Recommendation Q.1225, *Physical plane for intelligent network CS-2*, 1997

[4] ITU-T Recommendation Q.1211, *Introduction to intelligent network CS-1*, 1993

[5] ITU-T Recommendation Q.1229, *Intelligent network user's guide for capability set 2*, 1997

# Chapter 1.3

# BROADBAND SIGNALING

Signaling has been defined as 'the system which provides the ability to transfer information between customers, within networks and between customers and networks' [1]; and also as 'the exchange of control information between elements of a telecommunication network' [2]. The common denominator of all different definitions is that in a telecommunication network signaling plays the same role as the nervous system in a living organism [3]. The signaling system is in charge of the overall coordination of the network elements binding together to give a cohesive entity. Therefore, signaling comprises both the semantical and syntactical information exchange for control and supervision of the network. These mechanisms include both the call/connection control procedures for setting up, releasing and, in general, handling calls and all the operations dealing with the network management and maintenance.

From the very beginning of the old-fashion telecommunication switches until the modern digital networks, the signaling systems have evolved in parallel to the technological changes. The first signals were rudimentary dial pulses or distinct frequency combinations. In past systems, signaling information and user speech are conveyed together over the same channel giving rise to the term in-band signaling used to describe such systems. In modern telecommunication networks, signaling is considered as a data service transferred over a common data channel, like the D-channel of narrowband ISDN. This has been described as out-of-band signaling.

The wide-spread acceptance of the common channel signaling approach has led to the introduction of data networks for the transfer of signaling messages containing a large variety of information elements. The semantics of the information have been enriched and the transfer speed is increased, thus, permitting the support of a great variety of complex services.

The rise of B-ISDN has provided the platform for the introduction of multimedia services which effectively combine video, audio, data and other types of information (for a discussion of multimedia services and associated network technologies see Chapter 1.1). Although the target is a universal integrated services network or at least a set of inter-operable heterogeneous integrated services networks, each service presents distinct traffic features and Quality of Service requirements. Therefore, the signaling system should provide the means for handling the most complex service allowing at the same time the selection of more simple modes of service provisioning. Such a signaling system will be able to accommodate the pressing demand for optimized connection handling procedures, while guaranteeing the efficient operation of the Broadband network

*Intelligent Broadband Networks.* Edited by I. S. Venieris, H. Hussmann
© 1998 John Wiley & Sons Ltd.

as well as its backward compatibility to less advanced signaling systems. It is only natural that several approaches to the design of signaling systems exist, ranking from the most complex in which services are structured as several associations between elementary reusable objects representing not only physical resources but also basic object associations, to the most simple in which complex service requests are treated in dedicated servers of the network while the signaling system is restricted to a mere recognition of a service as a 'complex' one.

This chapter attempts to give a brief overview of Broadband signaling. As this is a quite general task, not all signaling architecture alternatives proposed for the future multimedia networks are covered. The main emphasis is put on standardized Broadband signaling systems currently incorporated in public switched Broadband networks. To ease the discussion the chapter starts with more general aspects relating to signaling and continues afterwards with the systems description not omitting to show the evolution from narrowband to Broadband networks.

## 1.3.1   Basic Principles

The traditional path that leads to the definition of the signaling capabilities is originated by the definition of the services from the user perspective followed by the identification of the requirements that the network must provide. These requirements address both the physical infrastructure (for example the bandwidth needed by the user applications and the switching techniques) and the resources for controlling and managing the physical infrastructure, that is the 'intelligence' of the network.

If one of the requirements imposed by the service is the presence of a switched network able to allocate the resources upon user request then the need for signaling capabilities is mandatory. For example, this requirement can be driven by the 'on demand' characteristics of a service and the related charging policy: if the service is invoked by the user only a few times during a given period of time, it is not feasible to force the user to make use of leased lines or semi-permanent channels unless the network provider offers very cheap rates.

In B-ISDN for example, the underlying ATM technique has been designed adopting a connection-oriented approach (see Section 1.1.5.3). Therefore it is natural to define a Control Plane for hosting the signaling system that undertakes the allocation, the modification and the release of the ATM network resources. These resources are typically identified as Virtual Path/Virtual Channel Connections (with the allocated bandwidth) and supported by mechanisms like policing in order to guarantee the network usage and to assure the Quality of Service. Therefore, the switched B-ISDN exploits not only the transport mechanisms offered by the User Plane but it is strictly linked to the signaling capabilities provided in the Control Plane.

The term signaling capability mainly encompasses two concepts: the requirement and the protocol. The requirement can be seen as the definition of the feature that must be realized by means of the signaling system of the network:

this definition is composed by the clear identification of which Functional and Physical Entities are affected by the introduction of the new feature and by the specification of the information and data, named as Information Flows, that must be exchanged between each pair of Functional Entities.

Accordingly, the signaling protocols define the set of actions and attributes across an interface needed to provide the services. Then, a signaling protocol includes both the syntax and the semantic of the exchanged information which constitute the external information flows between Functional Entities as explained above. Moreover, the signaling protocol determines the set of Protocol Data Units (PDUs) and State Machines for the communication between signaling application processes of the signaling system and uses the services of the transport system.

In the following chapters the different items briefly defined here are further elaborated. More precisely the discussion continues with a description of a simplified functional model for B-ISDN, a list of additional signaling capabilities required by multimedia services, and a description of the signaling protocols currently employed of the support of these requirements.

## 1.3.2   B-ISDN Functional Model

In narrowband ISDN (N-ISDN) the signaling system can be based on the principles of a simple telephone call between two parties involving a bearer connection. On the other hand B-ISDN requires a more elaborate signaling system based on a new definition of the call concept which implements the simultaneous handling of many parties by appropriately utilizing network resources.

These requirements are imposed by the expected evolution of services from the existing monomedia two-party services to future multimedia services involving complex party associations. Additionally the transmission/reception preferences may vary from party to party. Also the way by which the various media are presented maybe different depending on each party preferences or terminal capability. Moreover synchronization requirements may exist since each call is now composed of several different media. Plus the individual characteristics of each party and each bearer connection in a call are now time-varying (e.g. bearer connection bandwidth can be re-negotiated) and each call may also be modified by either splitting into new or joining with other calls.

The complex functionality of such a signaling system can be decomposed into simpler structures by means of Functional Entities (FEs) and Functional Entity Actions (FEAs). Their definitions are the same with the ones appearing in Section 1.2.2. The use of FEs makes the design more modular, adds support for a great deal of services while it also facilitates the network evolution. However such a grouping of functions does not preclude the way by which specificFEs are allocated to physical network components. It is obvious that the establishment of a B-ISDN call requires a clear separation between the FEs which control the establishment of associations between the parties and the network and the FEs which control the allocation of network resources required for the services. This is in contrast to the N-ISDN approach which groups the call control and bearer connection control functionality within the same entity. The former approach is called separate while the latter is called monolithic (see Figure 1.3.1).

In the separate approach the basic principle is the decomposition of a communication session into several different independent tasks. Therefore one could recognize at least the call control and bearer connection control tasks while other subdivisions like the bearer channel control task are not excluded. The motivation behind this separation is to distinguish among network elements offering subsets of signaling capabilites. For example in the case of the ATM Passive Optical Network described in [4] only channel control functions are required in the PON while higher level control functions are located to the Local Exchange and the user terminal. With this organization a simple to realize but also dynamically operating access network is achieved. The same approach can be efficiently applied to the case of wireless ATM access network as discussed in [5].

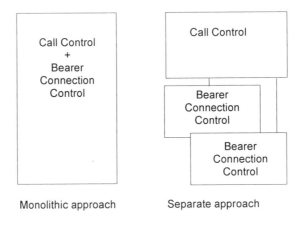

Figure 1.3.1
Monolithic and separate approaches

The signaling functional model for B-ISDN is described in [6], and uses the method specified in ITU Recommendation I.130 [7] for modeling telecommunication services. In particular the B-ISDN signaling model complies with the Stage 2 methodology. In short, ITU has standardized general modeling methods for characterizing telecommunication services. The method recognizes three main stages: Stage 1 includes the prose definition of a service and proceeds with a static and dynamic description of the service in terms of attributes and SDLs (Specification and Description Language). It expresses the user perception of a service and in this respect the description is implementation independent. Stage 2 attempts to define a functional model and the Information Flows among FEs as well as the FEAs referred to previously. It is also network independent in the sense that it does not describe how the proposed model will apply to a particular signaling system. Stage 3 includes the implementation specification as it covers all aspects a network deals with, that is the signaling protocols of the switching nodes.

For the discussion following, Figure 1.3.2 depicts a simplified version of the B-ISDN Signaling functional model with respect to the corresponding one depicted in Figure 3.1 of [6]. The Control Plane is decomposed into a number of signaling FEs which are then distinguished into Call Control related FEs and Bearer Control related FEs. This way a functional separation between call control level and connection control level is applied. Co-ordination among signaling FEs is performed by Application Control (AC) functions. Call Control includes the DCA, DC and EC FEs illustrated at the upper level of the Terminal Equipment (TE), the Serving Node and the Relay Node. Bearer Control includes the LCA and LC FEs residing at the lower level of the Physical Entities. The AC is a single FE incorporating Call Control functions (overall control of the call, service admission control, communications with connection control, charging information), Connection Control functions (overall control of the connection, control network resources, collection of traffic count information), gateway only functions, communications with Link and Edge/End FEs and communications with IN Functional Entities. The main functions of the Link Control FE are the control of the link association between two LC FEs and the transport of AC information between two adjacent AC FEs. The control of the edge control association between the EC FEs and the transport of AC information between two AC FEs located at the edge of the network are due to the EC FE.

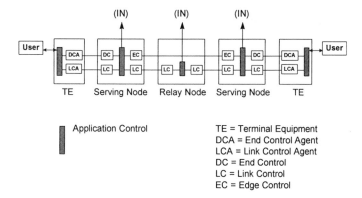

**Figure 1.3.2**
B-ISDN Signaling Functional Model

The Stage 2 functional model specifies also the information flows (IFs) between the FEs identified; the external IFs constitutes the basis for the definition of the signaling protocols.

## 1.3.3   Broadband Signaling Capabilities

If the introduction of signaling in B-ISDN was limited only to the establishment/release of a single connection call between two parties, the existing narrowband signaling systems could be easily reused. This is because the main modification required would be that the basic unit assigned/released by the network would be an ATM Virtual Channel Connection instead of a PCM Time Slot. This has a major impact on the interface between the signaling system and the switching node fabric, but it has a minor impact from the protocol point of view.

Since the support of switched virtual channel connections is a de facto characteristic of B-ISDN signaling, the current chapter deals with some of the most interesting additional signaling capabilities that have been considered by the standardization bodies.

*Asymmetric Bandwidth:* the interactive retrieval services described in Subsection 1.1.2.3 generally require a high bandwidth in the direction from the server to the user, while the bandwidth from the user to the server can be very small, as it is used only for sending selection commands. In this case it would be a waste of network resources to reserve the same bandwidth value for both directions. The signaling system must therefore allow the specification of different values for the forward and backward direction of a connection.

*Additional Traffic Parameters:* one of the advantages of the ATM technique is the capability to optimize the allocation of network resources by statistically multiplexing cells issued by different sources. To do so, the connection admission control mechanism should know the exact traffic characteristics of each source. However if the signaling messages transported only the Peak Cell Rate value of the source, the network would not be able to know if the source transmits at a steady rate or if bursts of cells interleaved by silence periods are coming in. To avoid bandwidth allocation based only on the Peak Cell Rate, additional parameters must be transported by the signaling messages and managed by the Application Process; the Sustainable Cell Rate and the Maximum Burst Size are two parameters that are the input for computing the expected bandwidth and for applying the policing algorithms in order to verify that during the active phase the source respects the declared values.

*Negotiation:* this capability allows the calling user to indicate during the connection establishment different bandwidth values that are acceptable for the invoked service.

*Modification:* this capability is oriented towards bandwidth modification of a connection during the active phase. For the time being, only the modification of the Peak Cell Rate of a point to point connection has been completely specified. This capability is useful for services where the calling user does not know in advance the bandwidth in the backward direction, that can therefore be assigned by the called user after the acceptance of the call.

*Point-to-Multipoint:* the distribution services described in Subsection 1.1.2.4 require a 'distribution tree' connection topology. Therefore ATM cell replication functions are required inside the switching fabric as well as new signaling features to control the creation/dropping of branches of the distribution tree; the signaling procedures vary according to the initiating party (root initiated or leaf initiated procedures).

*Multi-connection:* this is one of the capabilities that are better suited for the application of call and connection control separation. In the switched narrowband networks, the concepts of call and connection are often overlapping because one call cannot exist without an associated connection; when one user desires to make a phone call, the establishment of the call (defined as the object representing the control association between the parties) is simultaneous with the establishment of the connection (defined as the object representing the network resources requested by the service). Moving to a Broadband environment, if one multimedia service requires more than one bearer connection it is possible to think of a call establishment which is distinct with respect to the connection establishment and therefore it could make sense to have a call without connections.

*Look Ahead*: this procedure allows a network to perform called terminal availability and compatibility checking of network resources without any commitment. This is a capability that attempts to optimize network resource usage; usually, the resources are reserved link by link when the call establishment proceeds therefore if the called terminal is not available, for example because it is busy, this resource reservation proves useless. When requesting a Broadband connection, the reservation of resources can become a critical issue and would be useful to know 'in advance' if the connection can be successfully established. The Look-Ahead is a procedure that tests the status (availability and characteristics) of the called terminal and only if a positive acknowledgement is returned then the resource allocation starts.

## 1.3.4    Signaling Protocols

Signaling protocols for narrowband networks have evolved slowly along with terminals and network nodes. In the end N-ISDN adopted a message based signaling system, influenced by both telecommunications and data networks, which includes two protocols: DSS1 (Digital Subscriber Signaling No. 1) at the User Network Interface (UNI) and SS7 (Signaling System No. 7) at the Network Node Interface (NNI). Signaling architectures are aligned to the OSI Basic Reference Model (OSI BRM) principles which lead to protocols organized in layers.

The existence of two different protocols for the same kind of network resulted from the assumption that the network would always be much more intelligent than the user terminals. This assumption was a valid one having in mind plain telephone networks, nevertheless the evolution of telephone devices into PC-like terminals bridges the gap between terminal and network intelligence. Likewise DSS1 and SS7 become more similar as they evolve.

When moving to the design of the B-ISDN protocol architecture, the need existed for a trade off between the backward compatibility (that is the possibility of including the narrowband services in the Broadband network, allowing a dialogue between a Broadband terminal and a narrowband one) and the integration of new concepts coming from other environments (such as the OSI architecture). The UNI is now specified by DSS2 (Digital Subscriber Signaling No. 2, Appendix A, Tables A.5, A.8). An overview of the respective protocol stack is given in Figure 1.3.3:

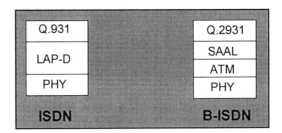

**Figure 1.3.3**
ISDN / B-ISDN UNI Protocol Stacks

*Q.2931* is the signaling protocol used for point-to-point call and connection control in Broadband ATM networks. Its messages are similar to those of Q.931, its narrowband counterpart, with additional information elements to support the enhanced capabilities required by multimedia services. For example, the SETUP message includes the *ATM user cell rate* information element which allows the specification of other than the Peak Cell Rate traffic parameter like the Sustainable Cell Rate and the Maximum Burst Size of the connection. These parameters are defined for both directions to and from the network so that bidirectional connections with asymmetric bandwidth can be established. To support point-to-multipoint connections a set of new messages is foreseen in Q.2971, the enhanced version of Q.2931 (see Appendix A, Table A.9). The ADDPARTY message enables the addition of new leaf nodes to an active connection between a user at the root of a logical tree and a user at an initial leaf of the tree.

*Signaling ATM Adaptation Layer (SAAL)* consists of two parts and provides all functions required for the transfer of call and connection control messages with ATM cells. The Common Part provides unassured information transfer and implements the segmentation/reassembly procedure of upper layer messages into/from ATM cells. The respective protocol is the ATM Adaptation Layer 5 (AAL5) which constitutes the most simple adaptation protocol of the AAL as it requires no additional control information in the segment header. The Service-Specific Part is further divided into the Service-Specific Connection-Oriented Protocol (SSCOP) and the Service-Specific Coordination Function (SSCF) which adapts the SSCOP service to the SSCF user needs. SSCOP provides reliable transfer of messages by implementing error correction of corrupted messages and flow control (Appendix A, Table A.7).

Concentrating on NNI signaling, an overview of the respective protocol stacks is presented in Figure 1.3.4 .

*Message Transfer Part (MTP)* consists of three levels and provides connectionless message transfer. MTP1 corresponds to the OSI Physical Layer and provides a bi-directional transmission path for signaling. MTP2 corresponds to the OSI Data Layer and performs error detection, correction and monitoring as well as flow control. MTP3 (Appendix A, Table A.7) realizes Signaling Message Handling and Signaling Network Management functions. Signaling Message

Handling is responsible for deciding whether a message that arrives from MTP2 is addressed to a different signaling point (thus it should be routed) or to itself (thus it should be delivered to the appropriate MTP3 user or function). Signaling Network Management is responsible for signaling network reconfiguration in case of signaling link or signaling point failures.

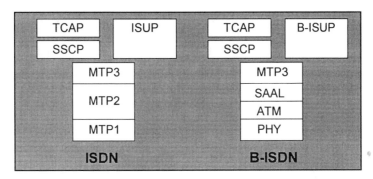

**Figure 1.3.4**
ISDN / B-ISDN NNI Protocol Stacks

*Signaling Connection Control Part (SCCP)* provides the functionality of the OSI Network Layer. Therefore it extends MTP3 addressing capabilities by being able to identify and deliver messages to SCCP users and also addresses messages with logical addresses (e.g. 800 numbers) and consequently transforms them to MTP3 parameters.

*ISDN User Part (ISUP)* provides call control functionality (establishment, supervision and release of a call) for voice and circuit switched data and video in ISDN. It also supports a number of supplementary services as for example calling-line identification, closed user groups, call forwarding, user-to-user signaling. *B-ISUP* (Appendix A, Tables A.4, A.6, A.9) is its B-ISDN counterpart which can additionally support high speed data applications, video services (e.g. Video Conferencing) and residential video distribution (e.g. Video on Demand) using ATM Virtual Channel Connections. In general, the B-ISUP messages are the same as those of ISUP while additional Information Elements allow an operation analogous to that of Q.2931. For example in the Initial Address Message of B-ISUP, an *ATM user traffic descriptor* IE is now present.

*Transaction Capabilities Application Part (TCAP)* realizes the dialog between applications residing at separate nodes through a query/response interaction. TCAP makes use of the services offered by SCCP and this is the reason for the inclusion of SCCP in the protocol stack for the NNI.

When dealing with signaling protocols usually the term is restricted to the protocols above the transport level, that is above layer 3. It was decided by standardization bodies to develop the signaling protocols above layer 3 trying to be as close as possible to the concepts of the OSI layer 7 (the Application layer). In particular the main interest was placed on the so called Application Layer Structure (ALS) which adopts an object-oriented philosophy in the representation of a communication session [8]. In ALS, two elementary blocks are introduced, namely the Application Entity (AE) and the Application Service Element (ASE). ASEs coordinated by the Single Associated Control Function (SACF) constitute a Single Association Object (SAO). An AE may consist of one or several SAO where each SAO is assigned the task to coordinate the association between different AEs. Overall the AE represents the specific communication actions of an Application Process which is defined as the set of resources of a system required to perform an information processing activity.

The ALS concept has been applied in both the Access Protocols (DSS2) and the Network Protocols (B-ISUP). An example can be found in Recommendation Q.2721 'Overview of B-ISDN Network Node Interface (NNI) Signaling Capability Set 2, Step 1' that is reported in Figure 1.3.5.

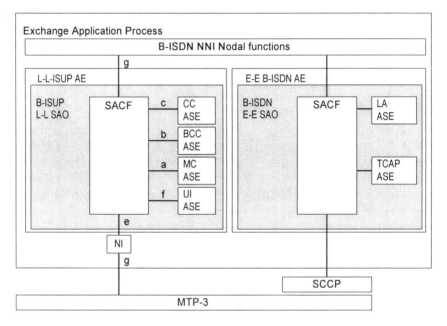

Abbreviations:

| | | | |
|---|---|---|---|
| AE | Application Entity | MC | Maintenance Control |
| SACF | Single Association Control Function | NI | Network Interface |
| ASE | Application Service Element | SAO | Single Association Object |
| BCC | Bearer Connection Control | UI | Unrecognized Information |
| LA | Look-Ahead | | |

**Figure 1.3.5**
B-ISDN NNI Protocol Architecture Model for CS-2.1
Reproduced with permission of ITU from recommendation Q.2721.1 full source of original publication.

It should be noted that the B-ISDN NNI Nodal functions (that represent the Application Process functions of the switch like routing, charging..) lie upon two Application Entities: the Link by Link B-ISUP AE is composed by the 'traditional' link by link protocol (in particular the CC ASE is devoted to the Call Control while the BCC ASE to the Bearer Connection Control) while the Edge to Edge B-ISDN AE includes the Look Ahead ASE that is the only capability that can cross transparently the transit nodes, making use of the services provided by the Transaction Capability Application Part (TCAP).

To further understand the flexibility of the ALS concept lets consider the case of ISUP in narrowband ISDN. As ISUP is monolithic both the call and bearer connection control functions will be grouped into the same ASE making it impossible to modify/enhance a set of functions without affecting the other. Also consider the case where separation does not only refer to a distinction of call and bearer connection control related messages in a single protocol but implies a complete separation of call and bearer control protocols. In this case the CC and BCC ASEs will reside in separate SAOs and the obvious advantage compared with that shown in Fig. 1.3.5 is the capability to establish various associations between a call and bearer connections. This possibility is not present in the case of Fig. 1.3.5 where the association between the call and the related connections cannot change.

The integration of the IN concept with B-ISDN is achieved by placing the Broadband Intelligent Application Protocol (B-INAP) over SCCP/TCAP. As described in Section 1.2.9, INAP handles the exchange of messages between the various IN Functional Entities by utilizing Application Service Elements (ASEs) which communicate through TCAP messages. A presentation of the complete protocol stack at the NNI, the enhancements which were introduced in B-ISDN for integration with IN as well as a detailed description of the B-INAP can be found in Chapter 2.3.

It is evident that the evolution of the signaling protocols is oriented towards the decomposition in small functional blocks, independent as much as possible with respect to the other blocks. Experimental signaling protocols (e.g. EXPANSE [9]) adopt an object oriented approach where each such functional block is realized as a simple object (referred to as elementary call object). A service can be constructed as a combination of elementary call objects which have been appropriately initialized. This way modifications imposed on an elementary call object can alter the characteristics of a service without affecting other components. More over new services or additional functionality within a service can be introduced by deriving new object classes from already existing classes of elementary objects. This approach guarantees a more flexible maintenance of the software of the digital exchanges if compared to the monolithic solutions where the upgrading of the capabilities is quite expensive as it involves a high portion of the code.

The view taken in this book mostly refers to existing signaling protocols and therefore the underlying assumption is that the signaling system is somehow unable to process calls for complex multimedia services. These calls are always transferred to the Intelligent Network and this is precisely how complex services are provided by a less sophisticated signaling system. To further show that the Intelligent Broadband Network architecture can easily extend to interoperate with advanced releases of signaling, two cases are considered: TINA in Chapter 3.2 and advanced Broadband signaling in Chapter 3.4.

## 1.3.5    Broadband Signaling Standardization

### *1.3.5.1 B-ISDN Signaling Capability Sets*

In the definition of the Broadband services there is in principle no limitation to the complexity of the scenarios. Topology with multiparty configurations linked by a variable number of connections can be easily considered for the provision of complex multimedia services. On the other hand, it is clearly impossible to foresee a sudden transition from an ATM backbone allowing only semi-permanent VP/VC connections to a switched B-ISDN network with a complete set of capabilities. It is therefore more reasonable to follow a phased approach in the introduction of the signaling in a Broadband network, starting from the simplest capabilities and increasing step by step the available features.

The ITU-T standardization bodies, responsible for the definition of the standards for public networks, have defined three main phases for the introduction of the signaling capabilities, as indicated in Figure 1.3.6. The ITU-T Study Group 13 is responsible for the definition of the general requirements of the B-ISDN, considering both User plane and Control plane characteristics. For example, one of the objectives of ITU-T SG 13 is the definition of the ATM Transfer Capabilities and the relative Quality of Service parameters. On the other hand the Study Group 11 (Switching and Signaling) must specify the Signaling requirements and protocols that are compliant with the general requirements defined by SG 13.

The ITU-T Study Group 13 has defined three Releases for B-ISDN. Release 1 is oriented to the introduction of simple (that is between only two users and with only one communication stream) services requiring Broadband. Release 2 aims at the introduction of the set of capabilities that should accomplish the flexibility required by multimedia services; for example, Release 2 should contain the features that allow:

- To manage point-to-multipoint configurations;
- To handle more than one connection stream between the involved parties;
- The possibility of negotiation and modifying the characteristics of the connections;
- To exploit the statistical multiplexing inside the ATM network according to the traffic characteristics of the sources.

Release 3 should complete the previous phases, adding the more complex features like multipoint-to-multipoint and every new capability that in the meantime the telecommunications market could request.

The terminology adopted by Study Group 11 is slightly different because the term Capability Set is currently used. If the Capability Set 1 of SG11 mainly corresponds to Release 1 of SG13 except some minor details, the Capability Set 2 only refers to a subset of Release 2 of SG13. In addition to that, the SG11 protocol groups' stepwise approach has generated the term step as a further decomposition of the Capability Set. This is the reason why the set of B-ISDN signaling protocols for Release 2 are named as Capability Set 2 step 1, step 2 and so on.

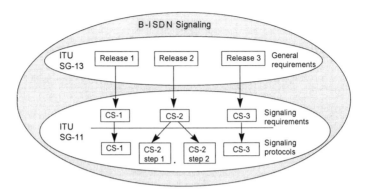

Release <-> ITU-T SG13
Capability Set <-> ITU-T SG11

**Figure 1.3.6**
Phases for the introduction of B-ISDN signaling

## 1.3.5.2 Status of Standards

The current status of the ITU-T specifications of the B-ISDN public network is described below:

- The entire Capability Set 1 has been approved in 1995;
- The Requirements of the Capability Set 2 has been finalized in 1994;
- The Protocols of the Capability Set 2 Step 1 have been approved at the beginning of 1997;
- Currently ITU-T SG11 is working on the Signaling Requirements for Capability Set 3, while Protocol Groups (DSS2 and B-ISUP) are working on Capability Set 2 step 2.

It is important to underline that in the last few years, new standardization bodies have addressed the signaling requirements and capabilities; this was caused by the emerging integration of Information Technology with traditional Telecommunication domains and by the emerging market needs.

The ATM FORUM is an organization born in USA and rapidly spread over the world; ATM FORUM's major objective is to define complete standards for the ATM. Regarding the signaling aspects, the ATM FORUM especially addressed the User Network Interface Protocols (the latest version is the UNI version 4.0), the interworking between different carriers (protocols for Broadband Inter Carrier Interface), the signaling of Private Nodes (protocols for Private Network Node Interface) and the Interworking Among ATM Networks (IAN group).

To avoid the proliferation of different standards, a harmonization activity with ITU-T standards has been pursued by both bodies.

DAVIC (Digital AudioVIsual Council) is another international organization devoted to the development of standards for multimedia services. It is due to DAVIC the introduction of the Session Control level, that is an additional level of service control that makes use of the capabilities provided by the usual call and connection control. Relations between session, call and connection are currently addressed by the Functional Model developed by ITU-T Capability Set 3 Requirements.

In the meantime the B-ISDN signaling should be able to consider requirements coming from Mobility standards, where control procedures not related to bearer connection are requested, for example in order to manage user service profiles.

Another major issue is the relationship between the traditionally connection oriented telecommunication networks (both narrowband and Broadband) and the more and more successful connectionless world driven by the Internet paradigm, based on TCP/IP protocols.

Finally the relationships between Broadband signaling and Intelligent Network are investigated and they are the main subjects addressed in Chapters 2.2 and 2.3 of this book.

This short scenario indicates that, as telecommunication technologies evolve, the work carried out in standardization bodies is becoming even more challenging. Nevertheless, in the same time, it is always difficult to cater for the emerging evolutionary paths since the reference scenarios are changing every few months.

# References

[1]  Manterfield R, *Common Channel Signaling*, IEE Telecommunications series, **26**, 1991

[2]  Schlanger G, *An overview of Signaling System No. 7*, IEEE J. Select. Areas Commun., **4**, 360-365, 1986

[3]  Modarressi A, Skoog R, *Signaling System No. 7: A Tutorial*, IEEE Commun. Mag., **28**, 19-35, July 1990.

[4]  Killat U ed., *Access to B-ISDN via PONs*, J. Wiley/Teubner, 1996

[5]  Loukas N, Passas N, Merakos L, Venieris I, *Design of Call Control Signaling in Wireless ATM Networks*, Proceeding of ICC 97, 1554-1559, Montreal, June 1997

[6]  ITU-T COM 11 R 35 *Broadband capability set 2 signaling requirements,* 1996

[7]  ITU-T Recommendation I.130, *Method for the characterization of telecommunication services supported by an ISDN and network capabilities for an ISDN*, 1992

[8]  ISO Recommendation DIS 9545, *Application Layer Structure*, 1990

[9]  Minzer S, *A Signaling Protocol for Complex Multimedia Services,* IEEE J. Select. Areas Commun., **9**, 1383-1394, 1991

# Part 2

## ARCHITECTURE OF INTELLIGENT BROADBAND NETWORKS: FUNCTIONAL AND PHYSICAL ENTITIES

# Chapter 2.1

# GENERAL FRAMEWORK FOR INTELLIGENT BROADBAND NETWORKS

Intelligent Broadband Networks enable high-quality multimedia services by combining the high bandwidth of Broadband (ATM) networks with additional functionality of the network, using concepts from the Intelligent Network architecture. However, an adequate support for multimedia services cannot be achieved by a simple transfer of the existing architecture of the Intelligent Network but some additional structure needs to be added.

In order to introduce the new architecture, this chapter first gives a motivation for the new features which have to be added to the IN architecture. Afterwards, the basic approach chosen for the enhancements is explained on a high level of abstraction.

## 2.1.1 Architectural Enhancements in Intelligent Broadband Networks

As it was explained in Chapter 1.1, advanced multimedia services have the following main requirements from the network architecture:
- Support of high bandwidth;
- Support of defined Quality of Service;
- Support of complex communication configurations (multipoint, multi-connection, multicast);
- Support of complex interaction between user and service;
- Negotiation of network resources and modification of network usage during runtime of a service.

An additional requirement is that a new architecture should provide the same advantages which were the motivation for the introduction of the Intelligent Network concepts. More precisely, this means:
- Deployment of new services and modifications of services without changing the core parts of the network infrastructure;
- High flexibility in billing for services;
- Generic availability of basic service features (like, for instance, call redirection) to all telecommunication services.

Some of the requirements listed above are met automatically by combining B-ISDN and IN. For instance, the bandwidth and Quality of Service requirements are met by properties of the B-ISDN network. However, there are two specific requirements which have a particular impact on the overall architecture of a combined network. These are the support of complex communication

*Intelligent Broadband Networks.* Edited by I. S. Venieris, H. Hussmann
© 1998 John Wiley & Sons Ltd.

configurations and the support of complex interaction between user and service. A brief discussion of these two topics follows below. At the end of the chapter, some guidance is given for the reader through the rather technical material of the following chapters.

### 2.1.1.1 Complex Communication Configurations

One critical requirement is the support of complex communication configurations. A service may comprise many separate connections, which belong to a single complex service like a Computer-Supported Co-operative Work (CSCW) application. For instance, there may be more than two participants in such a conference which need to be linked together, and several parallel connections may be needed in order to support different media types. The user perceives a multimedia service in a Broadband environment as a single unit, although the network may use a complex connection configuration. The architecture has to ensure that there is a single and homogeneous view of all pieces of a service. For example, billing for a CSCW service session should be based on the combination of all resources which were used instead of individual billing for the various connections. Now there are two basic options on how to proceed in a combination of B-ISDN and IN:

- The first alternative is to rely on advanced signaling features of the B-ISDN (Signaling Capability Set 3) in order to provide for the coordination of network resources belonging to a single service instance. This means that the notion of a 'call' is extended to multi-party, multi-connection calls and a service instance always corresponds to exactly one call.

- The second alternative is to provide a separate level of abstraction which provides exactly the required functionality for complex communication sessions.

In this book, the second alternative is pursued further. The reason for this decision is that the first alternative would require the availability of very advanced versions of Broadband signaling for which standardization has not yet reached a stable status and practical availability seems to be still far away. Moreover, the second alternative is the more general one, since it enables a migration from basic B-ISDN signaling (where a call corresponds to a single bearer connection) towards more advanced B-ISDN features without major changes in the multimedia services which rely on the more abstract session view.

The Intelligent Broadband Network, as it is defined here, introduces a detailed session model into the Intelligent Network. Figures 2.1.1 and 2.1.2 show the basic conceptual difference of the Broadband and narrowband Intelligent Network architectures. In Figure 2.1.1 the main Functional Entities involved in the control of an IN service are depicted. The state of the Call Control Function for a call is described by a Basic Call State Model (BCSM) which provides the view of a call which is available to the service logic. There is an application layer protocol which transports the necessary information to and from the Service Control Point in terms of the state changes in the BCSM. The basic assumption is here that a service instance always corresponds to a call.

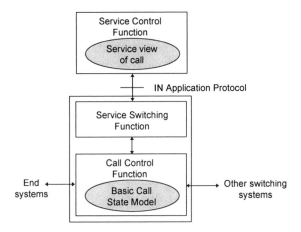

**Figure 2.1.1**
IN control in the traditional Intelligent Network architecture

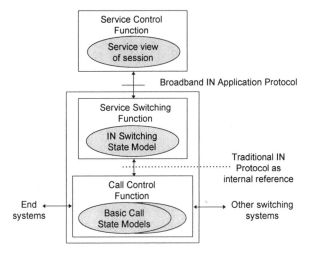

**Figure 2.1.2**
IN control in the Intelligent Broadband Network architecture

Figure 2.1.2 indicates the additional level of session control which is introduced in the Intelligent Broadband Network. In this architecture, the Service Switching Function (SSF) performs an abstraction and combination of information which may be based on several calls. The architecture still makes use of the Basic Call State Model but introduces an additional session state model on top of it. In order to be compliant with the traditional IN terminology, this book uses the name *IN Switching State Model (IN-SSM)* for the session state model. Of course, the application protocol towards the Service Control Function (SCF) is now formulated in terms of state changes in the more abstract IN-SSM, and therefore it is significantly different from the traditional protocol. However, the traditional protocol is still available as an interface between SSF and CCF, which is considered as an internal interface within the software running on a switching system. Nevertheless, the availability of this interface makes it easy to extend the architecture towards a 'hybrid' narrowband/Broadband architecture, just by making the traditional IN Application Protocol available externally, e.g. to combine a narrowband SCF with a Broadband SSF.

## 2.1.1.2 Complex User-Service Interaction

The second critical requirement, which was mentioned above, is the interaction between the user and a service that becomes much more complex in the case of Broadband network and multimedia services. The main difference to traditional Intelligent Networks is that here the end systems connected to the network are general-purpose computers rather than specialized devices like phone sets or fax machines.

A user connected to an Intelligent Broadband Network wants to use the full user interface capabilities of his or her computer for accessing multimedia services. For example, the access phase of a multimedia retrieval service should involve a highly interactive dialog where the user may browse through lists, watch preview clips of video files, and perform many other kinds of interaction. It is not possible to see these dialogs as part of the application software on the end systems. In fact, there are two kinds of dialogs: some dialogs are part of the application, while others are related to the network services and should therefore be available in a generic way for many different applications. Examples for the latter kind of dialogs are a generic Video on Demand gateway for many providers, a generic call redirection facility or a single directory of users for many different kinds of communication services.

The traditional Intelligent Network proposes to control user dialogs through a centralized service logic (executed on a Service Control Point, SCP) where the service logic can command a few specialized resources like acoustic announcements (executed by an Intelligent Peripheral, IP). This approach is not adequate for multimedia services, either, since it would lead to a very complex service logic and too complex communication between SCP and IP. Therefore, this book assumes that a special server is part of the Intelligent Broadband Network which carries out the more interactive parts of a service in a direct dialog with the user. This server, which is called a Broadband Intelligent Peripheral, is a general-purpose computer and runs its own kind of service logic in order to execute multimedia dialogs flexibly.

Besides this more complex kind of interaction between user and service, there is also the case where the interaction becomes simpler than in telephone-based services. For example, a user may invoke some service just by sending a message to the Intelligent Network, without having to establish a connection. This is a concept which has been studied already in Capability Set 2 of IN under the name of User-Service Information (USI) but which is not yet widely available. In the Intelligent Broadband Network, this concept is considered as a central requirement for the architecture from the beginning.

## 2.1.2   Overview of Part 2

The following chapters of part 2 give a detailed technical description of an Intelligent Broadband Network architecture. The basic structure is as follows: There are three chapters (2.2, 2.3, 2.4) which form the core part of the architecture definition, followed by three chapters (2.5, 2.6, 2.7) dealing in more detail with specific network elements of an Intelligent Broadband Network. Part 2 is concluded by a summary of the practical experiments, which were carried out in the INSIGNIA project (Chapter 2.8).

The first block of chapters (2.2, 2.3, and 2.4) should be of interest for any reader who wants to find out more about an Intelligent Broadband Network. In particular, Chapter 2.2 gives precise information on the functional architecture and therefore adds technical detail to the ideas, which were sketched above. Chapter 2.3 describes the new session-based IN Application Protocol. This description is carried to a significant level of detail in order to enable a thorough treatment of examples. Examples for the usage of the architecture and the protocol are presented in Chapter 2.4, including a more traditional network-oriented service (Virtual Private Network) as well as true multimedia services (Video on Demand, Video Conference).

The second block of chapters describes the network elements of a Broadband SSP (Chapter 2.5), Broadband SCP (Chapter 2.6) and Broadband Intelligent Peripheral (Chapter 2.7) in more detail. The information given here relates to the actual prototype implementations, which were developed in the INSIGNIA project and therefore provide a useful illustration for the general concepts introduced in the preceding chapters.

The final Chapter 2.8 gives an idea of the experiments, which were carried out to prove the correct function of the INSIGNIA prototypes. This is a report on practical experimentation with multimedia services on an international (native) ATM network in Europe, and therefore this chapter can be informative for an even larger readership than those interested in the architecture issues.

# Chapter 2.2

# FUNCTIONAL MODELS

The integration of IN and B-ISDN requires an evolution of the IN functional models which are defined in the IN Capability Sets 1 and 2 (see Chapter 1.2). The preceding Chapter 2.1 has given the main motivation for this evolution. Moreover, some basic ideas of the enhancements have already been explained there. This chapter now defines the functional models of an Intelligent Broadband Network in technical detail.

The reference functional model depicted in Figure 2.2.1 shows the main Functional Entities which are used to add intelligence to B-ISDN network services. The level of abstraction used in this chapter is the Distributed Functional Plane which deals with Functional Entities and does not yet take into account Physical Entities. The various Physical Entities of a Broadband Intelligent Network are discussed in other chapters of this part (2.5, 2.6 and 2.7).

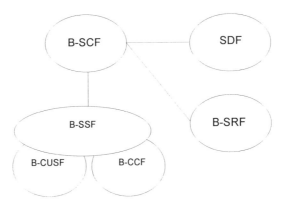

**Figure 2.2.1**
Functional entities of integrated IN/B-ISDN

The overall architecture shown in Figure 2.2.1 does not deviate from traditional IN, besides the fact that the Functional Entities are renamed individually into a Broadband version. The CUSF Functional Entity which is less widely known than the other entities has already been introduced in IN Capability Set 2 (see Section 1.2.3).

The Functional Entities depicted in Figure 2.2.1 are:
- Broadband Call Control Function (B-CCF);
- Broadband Call Unrelated Service Function (B-CUSF);
- Broadband Service Switching Function (B-SSF);

*Intelligent Broadband Networks.* Edited by I. S. Venieris, H. Hussmann
© 1998 John Wiley & Sons Ltd.

- Broadband Service Control Function (B-SCF);
- Service Data Function (SDF);
- Broadband Specialized Resource Function (B-SRF).

In this chapter particular emphasis is given to the Broadband Service Switching Function (Section 2.2.1), the Broadband Service Control Function (Section 2.2.2) and the Broadband Specialized Resource Function (Section 2.2.3). These three functional entities refer to a call model that is more powerful than the narrowband call model, as indicated in Section 2.2.4, and to the new IN concept of session that is presented in Section 2.2.5. How these new concepts are interrelated is explained in Section 2.2.6.

## 2.2.1   Broadband Service Switching Function

### 2.2.1.1 Definition of the B-SSF

The *Service Switching Function (SSF)* is defined in IN CS-2 as the set of functions which are required for interaction between call control and service control (see Chapter 1.2). This means that the 'hooks' in the call must be made available to the service logic residing in the Service Control Function. This general purpose of the SSF remains the same for the Broadband SSF. However, the detailed definition must be extended for the Broadband SSF considering innovative aspects. In the IN/B-ISDN functional architecture the role of the Broadband Service Switching Function is much more powerful if compared to the SSF in N-ISDN networks, mainly for the following reasons:

- The B-SSF must present to the Service Logic a much more complex configuration of network resources, considering that a service in a Broadband environment can request the establishment of many connections involving more than two parties. As it was explained in Chapter 2.1, a session concept is useful for this purpose;
- In order to give the IN full control over complex connection configurations, the B-SSF must perform actions on request of a third party which is the B-SCF. This means that call processing functions (in the B-CCF) are initiated on behalf of the B-SCF;
- The B-SSF must handle bearer-unrelated control channels between user and service logic, which means to deal with events that do not belong to any bearer connection.

In the remainder of this book, the term *session* will be used to denote an association of network resources (for example parties, bearer connections and bearer unrelated channels) for the realization of a single IN service or portion of a service.

From the Broadband Service Switching Point (B-SSP) point of view, the functionality required to provide an IN service can be mapped into one of the following three domains depicted in Figure 2.2.2:

- *Service Control Domain:* where the overall control of an IN service is carried out;

- *Session Control Domain:* where the overall control of the network resources associated to a session is performed;
- *Signaling Control Domain:* where the overall control of signaling events and switching resources associated to the B-ISDN calls/connections is realized.

In the Signaling Control Domain, the realization of a B-ISDN call is controlled by a component of B-ISDN switching which is traditionally named 'Call Control'. The Call Control handles both the control relationship and the physical switching resources needed for the communication between the involved parties. In addition to that the Signaling domain controls the resources used for handling events not strictly related to a particular call or connection (Call Unrelated Control, CUC). Call Control functionality belongs to the *Broadband Call Control Function (B-CCF)*. The Call Unrelated Control Domain is mapped on the *Broadband Call Unrelated Service Function (B-CUSF)*.

Consequently, both B-CCF and B-CUSF functionality pertain to the Signaling Control Domain. The functionality accomplished by the B-SSF belongs to the Session Control Domain, while that performed by the B-SCF pertains to the Service Control Domain. The next chapter introduces the main characteristics of the B-SSF. A deeper analysis of the relationship between the B-SSF and the B-CCF and between the B-SSF and the B-SCF is carried out in Sections 2.2.4, 2.2.5 and 2.2.6 below.

### 2.2.1.2 Characteristics of the B-SSF

The B-SSF functionality is located in the Broadband Service Switching Point Physical Entity. The B-SSP is the network element in which also the B-CCF and B-CUSF functionality are mapped, as shown in Figure 2.2.2.

As it was explained in Chapter 2.1, the most important enhancement of the B-SSF is the capability of handling sessions which may comprise several calls and call-unrelated communication relationships. The main part of this task is performed by the IN Switching Manager which works together with specific managers for calls and call-unrelated communication. The IN Switching Manager is the entity which maintains the central session model called IN Switching State Model (IN-SSM). The IN-SSM is explained in more detail in Section 2.2.5. It is one of the innovative aspects of the Broadband Intelligent Network architecture that the IN-SSM is an object-oriented model which can also be separated into two different views of an IN service, the overall *Party View* and a specific view of each connection *(Connection View)*. In this context, the term connection refers both to user-plane channels *(bearer connections)* and to control-plane channels *(bearer unrelated connections)*.

The CCF also keeps its own model on the level of a single call. This model is called a *Broadband Basic Call State Model (B-BCSM)* in analogy to traditional IN. In order to realize the session handling, the CCF must be enhanced such that it is able to coordinate several BCSMs. The same holds for the B-CUSF which holds the so-called *Broadband Basic Call Unrelated State Model (B-BCUSM)*. The B-BCSM and B-BCUSM are described in detail in Section 2.2.4 below. There is also another major difference in B-CUSF compared to IN CS-2 standards. The B-CUSF shown in Figure 2.2.2 does not have a direct interface towards the B-SCF, but it uses the capabilities provided by the B-SSF. This makes the whole architecture more orthogonal and enables an integration of call-unrelated relationships into the overall session model.

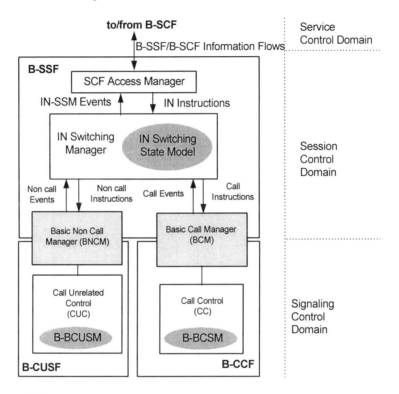

**Figure 2.2.2**
B-SSF/B-CCF/B-CUSF functional model

In the following, the functional blocks of Figure 2.2.2 are explained in more detail:

- The *Basic Call Manager (BCM)* is an extension of the BCM specified in ITU-T recommendation Q.1224 [1] and defined in Section 1.2.4. The major enhancement of the BCM is the handling of the Detection Point (DP) processing related to more than one side of the call (e.g. when SCP-initiated calls are considered). It determines whether a reached DP is armed, and whether it is armed as Trigger DP (TDP) or Event DP (EDP). The first encountered TDP originates the creation of a new session and of a new instance of IN service logic, whereas an EDP provides additional information for an existing session. The BCM functionality may be distributed among B-CCF and B-SSF, dependent on the actual B-SSP implementation;

- The *Basic Non Call Manager (BNCM)* defined here differs from the corresponding BNCM of recommendation Q.1224 [1] and defined in

Section 1.2.4, because it does not belong entirely to the CUSF, but is distributed between B-SSF and B-CUSF. Its role is to detect the call unrelated events (transported by a bearer unrelated connection) that must be reported to the IN Switching Manager and on the opposite direction it must address the service to user information coming from the B-SCF to the correct party;

- The Broadband Service Switching Function (B-SSF) is responsible for the realization of complex IN services on top of B-ISDN calls. The B-SSF comprises the *IN Switching Manager* and the *SCF Access Manager*.

— The IN Switching Manager keeps information on the status of a whole IN session in form of the object-oriented IN Switching State Model (IN-SSM), to coordinate several bearer connections and bearer unrelated connections belonging to the same session. The IN Switching Manager interacts with the SCF in the course of providing IN service features to users. It centers around the IN Switching State Model which provides a description of B-CCF call processing and B-CUSF call unrelated processing in terms of IN object states. The IN Switching Manager interprets the Call Events reported by the BCM and the Non Call Events reported by the BNCM and translates them into events describing state changes of the IN-SSM. These IN-SSM events are communicated to the B-SCF. In the reverse direction, IN instructions coming from the B-SCF are translated into instructions to the BCM or to the BNCM. The IN Switching Manager can also instruct the B-CCF to start call processing according to instructions from the B-SCF (SCP-initiated calls). Furthermore, the Switching Manager can instruct the B-CUSF to start a bearer unrelated connection processing for User Service Interaction.

— If IN processing is invoked (either by the B-SSF, on detection of an IN trigger, or by the B-SCF), communication between the B-SSF and the B-SCF starts. This communication is performed on the B-SSF side by the SCF Access Manager. The SCF Access Manager has the responsibility to locate the required IN service in the B-ISDN network and to send messages to and receive messages from the B-SCF. On the physical plane, this communication is performed using the *Broadband IN Application Protocol* (B-INAP). The main difference of B-INAP with respect to the standard INAP is that the contents of messages exchanged between B-SCF and B-SSF refer to the abstract IN Switching State Model and hide details of the Basic Call/Non Call Managers like DPs. For more details see Sections 2.2.4, 2.2.6 and Chapter 2.3.

## 2.2.1.3 Finite State Machines of the B-SSF

In the IN architecture it is common to further describe the functional blocks in terms of Finite State Machines (FSMs). Such machines give a rather clear idea of the behavior of Functional Entities in IN since the way how IN works can be seen as a translation between state machines of call processing and state machines of protocol handling.

For the design of a Broadband SSF, two distinct Finite State Machines (FSMs) are useful: one machine devoted to the communication with external entities (B-SCF), named *SCF Access Manager (AM) FSM*; a second FSM devoted to the SSM instance control which is called *IN Switching Manager (SM) FSM*.

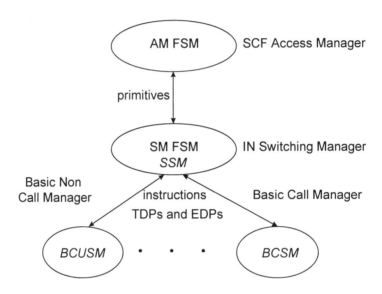

**Figure 2.2.3**
Finite State Machines inside the B-SSF/B-CCF/B-CUSF

Figure 2.2.3 shows the relationships between the identified FSMs and the Basic Non Call/Call Manager. Many Basic Call Managers or Basic Non Call Managers may be coordinated by a single IN Switching Manager FSM when more than one bearer connection or bearer unrelated connection is involved in the same service session. In any case, a single AM FSM for each SM FSM is in charge of the B-SSF/B-SCF communication.

## 2.2.2   Broadband Service Control Function

### 2.2.2.1 Definition of the B-SCF

In order to add value to multimedia Broadband services, the Intelligent Network must be able to exercise control on some functionality of the network. The Service Control Function (SCF) is the entity which is responsible for the network part of an implementation of a multimedia service. The SCF realizes different services by executing Service Logic Programs. The fulfillment of this role requires a simple and efficient IN interface between the Service Logic and the call/connection control function. The key principle for this interface is to choose an appropriate level of abstraction that hides the complexities of the network signaling interface but still provides the service logic with sufficient information to determine the appropriate action and instruct the call/connection functions appropriately. It has been discussed already above that it seems promising to use a session concept as the basis for this interface.

The main differences between narrowband SCF and Broadband SCF therefore can be found in the particular service logic and INAP protocol used between the SCF and SSF. The service logic contains specific control features for the Broadband network (e.g. session control) and these features are supported in an adapted B-INAP. However, the basic way how a Service Logic Program interacts with the SSF remains unchanged. The mechanism used to execute Service Logic Programs in the narrowband case is retained. The concept of interruption of call processing and interrogation of the SCF also remains identical, with the difference that 'call' processing now takes place on the level of sessions rather than individual connections. Generally, the overall architecture of the SCF remains identical to the narrowband case. This is, of course, an important advantage of the Broadband Intelligent Network architecture, since it allows the reuse of narrowband equipment and software for the Broadband case as well as a hybrid SCF which supports narrowband and Broadband services simultaneously.

As a consequence, everything which was said in Chapter 1.2 about the SCF remains valid for the Broadband case. In particular, Figure 1.2.11 above gives a decomposition of the B-SCF into functional blocks, the most important of which are the *Service Logic Execution Environment (SLEE)*, the *Service Logic Program Library (SLP Library)* and the *Functional Entity Access Manager (FEAM)*. Figure 2.2.4 recalls this structure in a more abstract way for the Broadband SCF.

In Figure 2.2.4, it is moreover indicated which internal models are necessary for the Broadband SCF. Conceptually, each Service Logic Program Instance maintains its own view of the configuration of a session. This model, however, is a 'mirror' of the IN Switching State Model for the respective session in the B-SSF. The B-INAP has been designed carefully for keeping the 'mirror view' consistent with the original IN-SSM instance.

From Figure 2.2.4 it can be seen that the B-SCF maintains interfaces with other Functional Entities besides the B-SSF. One of these Functional Entities is the Broadband Specialized Resource Function (B-SRF) which realizes additional resources for the execution of multimedia services (see Chapter 2.2.3). The other interface is towards the Service Data Function (SDF) which realizes a database for the storage of service-related instance data, for instance user names, user profiles etc. In this book, there is no explicit treatment of the SDF and its interface. Instead, the SDF is seen as an auxiliary database component for the SCF which is always located on the same physical machine as the SCF (i.e. on a Service Control Point). However, most of the considerations in the IN standards regarding the SDF can be obviously applied to the Broadband case in analogy.

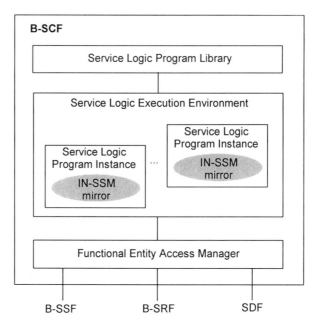

**Figure 2.2.4**
B-SCF functional model

## 2.2.2.2 Finite State Machines of the B-SCF

As for the B-SSF, it may be useful to consider briefly the Finite State Machines (FSMs) which are employed. The B-SCF FSM structure can be derived from the structure described in ITU-T Recommendations for IN CS-1 [9]. The basic FSMs which are required in a B-SCF and their relationships are shown in Figure 2.2.5.

The B-SCF contains the following FSMs:

- The Functional Entity Access Manager (FEAM) establishes and maintains the interfaces to the SSF, SRF and SDF. It passes/formats (and queues when necessary) the messages from/to the SSF, SRF and SDF to/from the SCME. In this way the SCF Management Entity (SCME) is relieved of low-level interface functions.

- The SCF Management Entity (SCME) comprises the SCME-Control and multiple instances of SCME FSMs. The SCME-Control is responsible for the creation, invocation and maintenance of the SCF FSM objects. It maintains the dialogs with the SSF, SRF and SDF on behalf of all the different SCF Session State Models (SSSMs). It interfaces the different SSSMs and the FEAM. For each context of related operations (session) an SCME FSM is created.

- The SCF Session State Model (SSSM). Each SSSM is associated with a service instance. It maintains dialogs with the SSF, SRF (and SDF) on behalf of this single Service Logic Program instance. This model corresponds to the SCF Call State Model from IN CS-1.

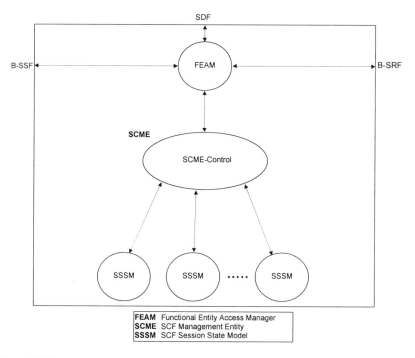

**Figure 2.2.5**
Finite State Machines inside the B-SCF

The detailed description of the main SCME FSMs, the SCSM FSM and the corresponding operations and procedures is omitted here. Further details can be found in ITU-T IN CS-1 Recommendations [9].

## 2.2.3 Broadband Specialized Resource Function

### 2.2.3.1 Definition of the Broadband SRF

The definition of a Specialized Resource Function (SRF) given in ITU Recommendation Q.1224 [1] is rather generic, but nevertheless strongly oriented towards narrowband networks and services (see Figure 1.2.12 in Chapter 1.2). The resources are mainly special hardware devices and user communication is done via narrowband audio channels. The transition to Broadband services has some significant impact here. Broadband services will need a new set of resources: user communication for Broadband services will be based on graphical user interfaces (GUI) that require a larger bandwidth, download of components into the Customer Premises Equipment (CPE) and frequent updates.

These services will utilize large databases for software components and additional service information. They will be based on multi-point connections, multi-connections and may even associate multiple sessions. These requirements result in the need for software resources. While an SRF managing hardware resources maps well onto the functional model presented in Chapter 1.2, an SRF for advanced software resources requires a refinement of the model. One possibility of a Broadband SRF architecture is depicted in Figure 2.2.6.

The SRF architecture shown here is very similar to the architecture of an SCF (see Figure 1.2.11 of Chapter 1.2). Many terms (e.g. Service Logic Library, SLEE) are identical to functional blocks of the SCF architecture. The idea is here to deal with specialized software resources in the same style as with Service Logic Programs. This can also be understood as a distribution of service logic over the B-SCF and the B-SRF functional entities. Of course, the B-SRF has its own particular kind of service logic which centers around advanced user interaction. The identical names just indicate the close similarity in architecture between B-SCF and B-SRF.

The components of this functional model that differ from the narrowband standards are characterized in the following paragraphs:

*Functional Entity Access Manager (FEAM)*. The FEAM is broken down into four blocks:

- The *SCF Access* manages the relationship with the SCF. It provides a reliable, sequenced message transport with request/response correlation;
- The *SSF Access* supports the relationship with the SSF and handles bearer connection setup and tear-down for bearer connections towards CPEs. This is a standard UNI signaling interface;
- The *User Access* provides reliable, sequenced message transfer with request/response correlation and support for asynchronous notification via bearer connections to the user. Alternatively stream oriented reliable transfer may be supported on a configurable per-service basis;
- The *Session Manager (SM)* provides the other functional entities with the abstraction of a session. An SRF session usually corresponds to a session on the SSF or SCF. A session bundles the relationships to the SCF and bearer connections. Higher abstractions that collect sessions together are not provided by the session manager. The session manager provides information about the sessions and their state to the other entities in the Service Logic and Instance Database. It also provides a mapping between the interfaces of the access modules and the Service Logic Execution Environment.

The *Service Logic and Instance Database (SLIDB)* holds generic, service independent information about sessions and service logic instances. It can be used by Service Logic Programs to associate multiple sessions into one service invocation.

The *Service Logic Library (SLL)* holds Service Logic Programs (software resources) ready for execution by the SLEE. These programs may be directly executable or may be interpretable scripts like HTML pages.

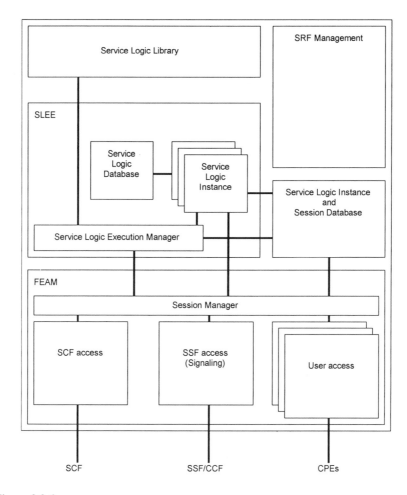

**Figure 2.2.6**
Broadband SRF functional model

The *Service Logic Execution Environment (SLEE)* is responsible for the execution of Service Logic Programs. It manages the invocation and execution of multiple Service Logic Programs and the concurrent access to other entities. It consists of the following components:

- The *Service Logic Execution Manager (SLEM)* creates service logic instances when necessary (usually if an incoming call for a new bearer connection towards a CPE is detected on the interface to the SSF). It selects the program to execute from the SLL, updates the information in the SLIDB and starts the program. Selection is done according to information provided in the event that triggers the creation and may be configured through SRF management;

- The *Service Logic Database (SLDB)* holds service dependent data, that may be either short-lived (only during the execution of a particular service logic instance) or long-lived (during the entire life-time of the service on the SRF). An example of the former is information about graphical objects in a particular invocation of a white-board service; an example of the latter are conference records in the Broadband Video Conference (B-VC) service. The database must support a locking scheme to regulate access collisions to the same data from multiple instances;

- The *Service Logic Instances*. A service logic instance is the invocation of a Service Logic Program for a particular session. Its lifetime spans from the creation of the session until its destruction.

The *SRF Management* is used to change the availability and configuration of the software resources. It manages a database of configuration data and interacts with all other functional entities.

## 2.2.3.2 Data Model of the B-SRF

The B-SRF, as all other Functional Entities handles the abstraction level of sessions. All information about the sessions is held in the Service Logic and Instance Database, which is accessible to the service logic instances and the service logic execution manager. The information which is kept here is a central data model for the B-SRF which can be structured into objects which belong to the following classes:

- *Services*. Each possible service consists of a number of software resources, which are implemented as Service Logic Programs, and is described by a static entry in the database. These entries are managed through management operations;
- *Sessions*. The session abstraction is used to bundle together all objects that belong to a service invocation. In the case of multi-party services there may exists an even higher abstraction, that associates together a number of sessions, but this should done by the Service Logic Program for this service. Each session belongs to a service;

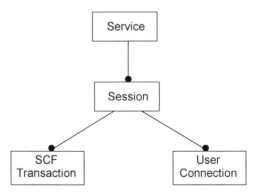

**Figure 2.2.7**
Class diagram for the FEAM

- *SCF-transactions.* Each session has one or more transactions with the SCF. These are usually implemented as TCAP transactions. Each SCF-transaction belongs to exactly one session;
- *User connections.* Each session has one or more bearer connections to users. If the last connection to a user is released, the session is destroyed. Each user connection belongs to exactly one session;

The relationships of these objects are depicted in Figure 2.2.7 using the OMT notation (see Appendix B).

## 2.2.3.3 User Connection State Machine

Instead of giving a complete account of all Finite State Machines used in the B-SRF (as it was done above for other Functional Entities), it is more informative to point out one specific state machine which is typical for the B-SRF. This state machine is required since connections between the terminals and the SRF are established by the B-SSF on behalf of the B-SCF, such that the B-SRF has to coordinate the dialog with the service logic (on the B-SCF) with the events taking place on its own signaling interface (where the SCP-initiated connection arrives as an incoming call). When the SRF receives a new connection, two things are to be synchronized:

- The SRF has to request instructions from the SCF. This must be done because the SRF does not know which resource the new connection belongs to. Depending on the implementation the FEAM may already know or not which service and session the new connection belongs to. The necessary information may be carried either in the signaling protocols or on the SCF transaction;
- The SRF must establish the higher protocol layers with the CPE.

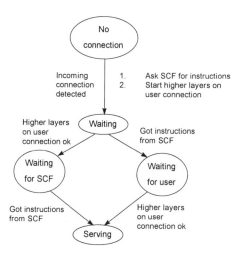

**Figure 2.2.8**
State model for a user connection

The Service Logic can start its execution only after these two events. This is realized by the State Model that is executed for each user connection. This model is shown in Figure 2.2.8.

## 2.2.4    Call State Models

### 2.2.4.1 Call Modeling in an IN/B-ISDN environment

The Call Control Function in the IN Distributed Functional Plane architecture models the activities of the switching/transport network underlying the IN. The behavior of the CCF is described by means of an abstract model called Basic Call State Model (BCSM). According to the IN recommendations [1], the BCSM is a high-level finite state machine description of CCF activities required to establish and maintain communication paths for users.

The IN CS-1 and CS-2 are targeted to the POTS and N-ISDN network, which provide essentially the traditional two-party call. Therefore the call models in the current recommendations have been designed to match the capabilities offered by the POTS and N-ISDN network. The call modeling must evolve to cope with the enhanced transport and signaling capabilities offered by a B-ISDN network. It is remarkable that the different releases of the B-ISDN standards, starting from the defined Signaling Capability Sets (SCS-1 and SCS-2.2) towards the future SCS-2.2 and SCS-3 impose very different requirements to IN 'call' modeling. In this chapter, a call modeling is described which fits to the available capabilities in the transport network, as it was used in the INSIGNIA project for carrying out field trials with existing switching systems. The contents of Section 2.2.4 is rather technical and specialized on call processing; so for a reader who wants to get an overview picture it may be preferable to skip to Section 2.2.5.

It is worth recalling the available network capabilities which provide the basis for the modeling choices. The B-ISDN SCS-1 has been taken as the starting point; some capabilities have been selected from SCS-2.2 (like a subset of the Generic Functional Protocol and of Bandwidth Modification procedures), while some new capabilities have been introduced (mainly the SCP-initiated call setup). The requirements on the call modeling can be derived as shown below:

- The use of SCS-1 call/connection leads to a Basic Call State Model (BCSM) based on the well known IN CS-1 model. The main enhancements concern the handling of this BCSM in the context of an SCP-initiated call. The goal is to define a call modeling that integrates the user-initiated and the SCP-initiated call in a coherent design;
- The use of Generic Functional Protocol leads to the introduction of the Basic Call Unrelated State Model (BCUSM), which has been derived from the existing IN CS-2 BCUSM.

The concept of call and connection separation is not needed in this context and a 'monolithic' approach has proved to be powerful enough. Advanced B-ISDN capabilities beyond the ones used in the trials have been taken into account from a more theoretical point of view. Some of these capabilities are: point-to-multipoint call/connections (belonging to SCS-2.1), call and connection separation and multi-connection (belonging to SCS-2.2 and beyond). In this section the models which have found an implementation in the trials of the INSIGNIA project will be described first, and then some considerations will be given on the possible future evolution.

## 2.2.4.2 Basic Call Modeling

The following description addresses the model for user-initiated calls first, and afterwards modeling of SCP-initiated calls is analyzed.

### Modeling user-initiated calls

As in the traditional IN approach, the call modeling for a user-initiated call foresees one BCSM instance for the originating side of the call and one other BCSM instance for the terminating side, within a single SSF. Each BCSM can be controlled by an independent service logic instance handling respectively an 'originating' IN service (e.g. a number translation service) and a 'terminating' IN service (e.g. call forwarding on behalf of the called user). This is known in the IN terminology as the 'functional separation' of the two halves of the call.

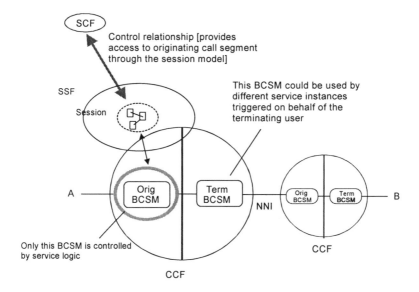

**Figure 2.2.9**
User-initiated call and BCSM instances

Hereafter only services interacting with the originating side are taken into account. This is not an architectural restriction: in principle the architecture supports also the 'terminating' control, but the services chosen for the demonstration do not require it. Figure 2.2.9 shows the modeling of a user-initiated call. This scenario is the same as in the IN CS-1 and CS-2 services, with the introduction of the session above the leftmost originating BCSM. In the picture the user B is connected to a remote SSP where two additional independent BCSMs could be instanced.

### Modeling of SCP-initiated calls

The *SCP-initiated call* is a core feature for the Intelligent Broadband Network architecture. The SCP can give instructions to the SSP to establish and control a connection between two users (it is a particular case of the more general 'third party call setup'). The BCSM modeling becomes more complex due to the need to monitor independent events coming from the two involved end systems. The requirements for the IN modeling of SCP-initiated calls are:

- The independent control of originating and terminating aspect, keeping the half call approach;
- The independent control of the activities of the two parties. For example the connection may be successfully established only towards one of the two parties, while the other party does not answer or is busy.

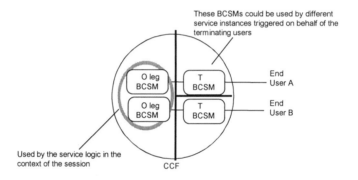

**Figure 2.2.10**
SCP-initiated call and BCSM instances

To fulfill the first requirement, SCP-initiated calls will be controlled at the originating side, allowing two independent terminating BCSMs at the terminating side. To satisfy the second requirement, a single state machine on the originating side becomes very complex. This is due to the inherent parallelism of the two connections towards the called parties. The solution is to instance two originating 'leg' BCSMs for the control of the two called parties, as shown in Figure 2.2.10. The structure of these leg BCSMs is exactly the same as for the user-initiated BCSM. The two leg BCSMs at the left side represent the call processing activities monitored by the service logic that requested the SCP-initiated call. These activities are resembled as originating activities, in fact the Call Processing inside the SSF/CCF has to perform operations similar to those performed in a standard originating side, e.g. Authorization, Analysis and Route Selection. For each called user the terminating activities can be monitored by the independent BCSMs shown in the right side of the picture. The control of the terminating BCSMs is not needed for all services. A reasonably powerful set of services can be realized without this feature, as demonstrated by the INSIGNIA project which did not implement terminating BCSMs. Additional aspects of the SCP-initiated call modeling will be analyzed later in this chapter.

### Leg BCSM

The diagram of the BCSM is shown in Figure 2.2.11. It is called *leg BCSM* to underline that it will be used to represent the originating side of a user-initiated call as well as each leg of an SCP-initiated call. The DPs which were actually used in the INSIGNIA project to realize the example services of the project are marked in gray. A brief description of each Point In Call (PIC) is given hereafter.

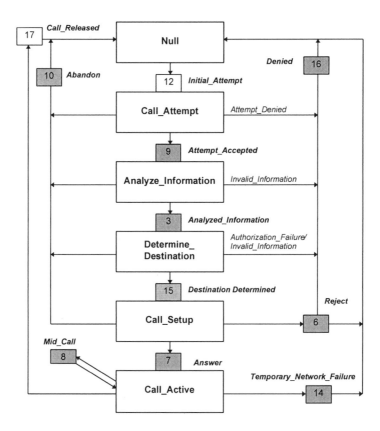

**Figure 2.2.11**
Basic Call State Model

*1.     Null*
The entry event is the disconnection and clearing of a previous call (DPs: Call_Released, Abandon, Reject, Denied and Temporary Network Failure). In this PIC no call relation exists, supervision is being provided. The exit event is the indication of an attempt to establish a call, either by the reception of a Setup/IAM message or by a command of the SSF (DP: Initial_Attempt).

*2.    Call_Attempt*

The entry event is the indication that the originating party needs to be authorized (DP: Initial_Attempt). In this PIC the rights of the originating party to place the call with given properties (e.g. bearer capability) are checked using the party's identity and profile. The exit events are the successful authorization (DP: Attempt_Accepted); the disconnect indication received from the originating party (DP: Abandon); the unsuccessful authorization (DP: Denied).

*3.    Analyze_Information*

The entry event is the indication that the originating party is authorized (DP: Attempt_Accepted). In this PIC, an initial information package is collected from the originating party. As the en-bloc signaling method is adopted, the information string is verified as a whole in order to check its correctness. The exit events are: the information string is correct (DP: Analyzed_Information); a disconnection indication is received from the originating party (DP: Abandon); the information provided by the originating party is invalid (DP: Denied).

*4.    Determine_Destination*

The entry event is the availability of a checked information string (DP: Analysed_Information). In this PIC the information is analyzed and/or translated according to the numbering plan to determine routing address and call type and the authorization of the originating party to place this particular call is verified. The exit events are: the Called Party destination (routing address) is determined and call setup is authorized (DP: Destination_Determined); the originating party abandons call (DP: Abandon); the call setup is not authorized because there are no rights to place the call (e.g. toll restricted calling line) or the information is not sufficient to perform the routing (DP: Denied).

*5.    Call_Setup*

The entry event is the availability of an authorized network address for the Called Party (DP: Destination_Determined). In this PIC the call setup is performed. The signaling protocols are used in the underlying network to establish both the call and the connection towards the called user (see ITU-T Q.2931 [3] for UNI and Q.2761-4 for NNI [4-7]). The exit events are: an indication that the call is rejected is received, the reason for the rejection is indicated in the CAUSE-value in the received B-ISDN signaling message (DP: Reject); an answer indication is received from the called party (DP: Answer); a disconnection indication is received from the originating party (DP: Abandon).

*6.    Call_Active*

The entry event is the indication of the answer of the called party (DP: Answer). In this PIC call supervision is being provided. The exit events are: a service/service feature request (e.g. for bandwidth modification) is received from one of the involved parties (DP: Mid_Call); a disconnect indication is received from one of the involved parties (DP: Call_Released); a general failure occurs (DP: Temporary_Network_Failure).

***Additional aspects of the control of SCP-initiated calls***

The features of the SCP-initiated call as provided by a network node can be very different according to the desired level of control, flexibility and performance. For example, it is possible to establish the two branches of the call in a parallel way or in a sequential way; in the second case this means that one leg BCSM can reach the PIC Active while the other leg BCSM is yet in Null PIC. Another problem is how to handle the failure in the establishment of one of the two branches (e.g. one called party refuses the call or does not answer). A sophisticated solution would be to report the failure to the service logic and, if needed, try to reroute the call, while a simple solution is to tear down the whole SCP-initiated call. As far as the INSIGNIA trials are concerned, neither the parallel nor the sequential call establishment is prescribed and the 'rerouting' capability is not needed, so that an SCP-initiated call can be torn down if one branch fails.

The model depicted in Figure 2.2.11 gives a high level of detail to control an SCP-initiated call. The status of the SCP-initiated call is represented by the combination of the two leg BCSMs. Given an arbitrary PIC x of leg BCSM A and an arbitrary PIC y of leg BCSM B, the overall status should be represented by the couple {x,y}, but four phases can be envisaged in an SCP-initiated call:

- *Establishment phase* - The leg BCSM A can be in PICs: Null, Call_Attempt, Analyse_Information, Determine_Destination and Call_Setup. The leg BCSM A can also be in the Call_Active PIC, given that the leg BCSM B is not in the Call_Active PIC. The leg BCSM B can be in PICs: Null, Call_Attempt, Analyse_Information, Determine_Destination and Call_Setup. The leg BCSM B can also be in Call_Active PIC, given that the leg BCSM A is not in the PIC Call_Active.
- *Unsuccessful establishment phase* - Both leg BCSM come back to Null PIC and at least one of the two leg BCSM is not making the transition to Null PIC coming from PIC Call_Active.
- *Active phase* - Both leg BCSM A and B are in PIC Call_Active.
- *Release phase* - Starting from the active phase, at least one leg BCSM has performed the transition from Call_Active to Null PIC.

If the service logic is interested in the monitoring of each single event in the two sides of the call/connection, all the DPs coming from the two BCSMs will be reported to the B-SSF. On the other hand, the B-SSF may request operations on the entire call and may prefer to have a more compact view of the processing of an SCP-initiated call. In this case a coordination of the two sides at the B-CCF level is required. A possible solution is to describe this coordination by means of a set of rules, which allow to report only a subset of the encountered DPs to the SSF. In Figure 2.2.12 the set of rules is shown as a sort of filter between the BCSMs processing and the SSF. The processing of the set of rules can be performed in the Basic Call Manager.

The set of rules for the coordination of leg BCSMs will be given below. It is important to underline that the application of this set of rules at the CCF level is not the only possible solution. It could be possible to coordinate leg events at the IN Switching Manager level. Moreover the proposed rules match the capabilities which are available in the switches for the support of SCP-initiated calls. Therefore, no rerouting is allowed, while it is assumed that the B-SSF can ask the B-CCF for the notification that a whole call has been established and that an active call has been released. Other rules may be defined in a future extension when more advanced capabilities will be available from the underlying network.

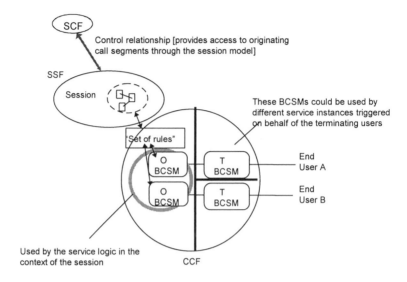

**Figure 2.2.12**
B-CCF coordination of leg BCSM

*Transition to Active phase*

**Rule 1**: the B-SSF asks for the notification that a call is in Active phase. The DP7 (Answer) of the two legs are armed and the following rule applies: the B-SSF is notified only when the second DP7 trigger is received.

**Rule 2**: the B-SSF asks for the notification when a loop from 'Active phase to Active phase' happens, that is when one trigger DP8 is received. The following rule holds: the B-SSF is notified when the first trigger (from one of the two leg BCSMs) is received.

*Transition to Release phase*

**Rule 3**: the B-SSF asks for the notification of a call in Release phase when one trigger DP9 or DP14 is received The following rule applies: the Basic Call Manager considers the whole connection released if one of the parties releases the call. In this case the notification to the B-SSF is performed when the first trigger (from one of the two leg BCSMs) is received. This rule is aligned with the underlying behavior of the B-CCF that is: a release from one side of the call is always forwarded immediately to the other side of the call.

*Transition to Unsuccessful establishment phase*

**Rule 4**: a call is considered as unsuccessfully established if one of DP 6 or DP 16 is notified by one of the two Leg BCSMs. The other Leg BCSM is immediately informed and the call towards the other side is released. No notification from this other leg BCSM is forwarded to the B-SSF.

It is worth mentioning that to allow a clear representation of the operations on the SCP-initiated call as a whole, the described call modeling could be enhanced.

A third 'master' BCSM could be superimposed to the two leg BCSMs in order to provide a more compact representation. In this case the requests coming from the SSF and the event reports coming from the CCF could directly refer to PICs and DPs of the 'master' BCSM. The master BCSM would replace the set of rules shown in Figure 2.2.12.

## 2.2.4.3 Basic Call Unrelated Modeling

The use of a bearer independent signaling mechanism at the UNI seems very promising allowing for example to support User-Service Interaction. A Basic Call Unrelated State Model (BCUSM) is therefore needed in the Intelligent Broadband Network architecture. According to the IN CS-2 architecture, the Call Unrelated Service Function (CUSF) provides the means to control the out-channel interactions with users and contains the BCUSM. The main difference to the IN CS-2 is that the BCUSM is not a view of call unrelated processing directly offered to SCF. The B-SSF interacts with the B-CUSF offering the SCF a view of bearer unrelated connection processing in the context of the SSM model.

It is useful to sum up the main assumptions for the signaling features that will be used in the following model for out-band interaction. A subset of the Generic Functional Protocol as standardized in Q.2932.1 [8] will be used at the UNI; therefore only local mechanism will be used, i.e. the signaling relationship will be terminated at the UNI between the user and the local exchange. The *Connection Oriented Bearer Independent (COBI)* mechanism will be used but the COBI relationship will be always setup in the user to network direction. The BCUSM is a modification of IN CS-2 as shown in Figure 2.2.13.

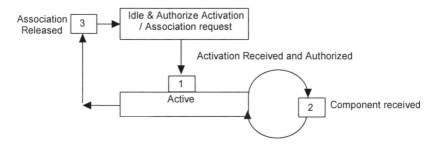

**Figure 2.2.13**
Broadband Basic Call Unrelated State Model (B-BCUSM)

Compared to IN CS-2, a new DP 'Association Released' has been added. It will be used by the service logic to monitor the release of the association. The 'Association Release Requested' DP and the 'Release' PIA (Point in Association) of the IN CS-2 BCUSM have been removed. A description of the states of the models is given below. The name *Point in Association (PIA)* is used here for the states of a BCUSM (in accordance to IN CS-2).

*1.   Idle and Authorize Activation / Association Request*
The entry event is the clearance of a previous association. In this PIA no

association exists (no call reference is assigned), when a request to establish a COBI association is received (i.e. COBI-setup message received from the user), the rights of the originating party are checked. The exit event is the request to establish a COBI association (and optionally invoke an operation) received from the user (DP Activation Received and Authorized).

*2.    Active*

The entry events are: a request to establish a COBI association (and optionally invoke an operation) received from the user (DP Activation Received and Authorized) or a component is received from the user (DP Component Received). In this PIA an association exists (a call reference is assigned) and the received operations are processed. The exit events are: a component is received from the user (DP Component Received); an association release request is received by the user or by an SSF instruction and the association is released (DP Association Released).

## 2.2.5    Session State Model

### *2.2.5.1 The Session Concept*

The objective of IN and B-ISDN integration is to provide complex services on top of the ATM network infrastructure. The set of B-ISDN basic services is not fixed, but it follows an evolutionary path according to the successive signaling capability sets. Therefore, a requirement to any architectural solution is to be suitable for short term implementations as well as easy to upgrade.

As a single IN service can be composed of a complex set of network resources, the modeling of this aspect requires the introduction of a new concept for IN, the concept of a *session*.

In the telecommunication world the concept of session is widely used but it has different meanings according to the context where it is applied. In the IN/B-ISDN integrated architecture, the concept of Session Domain has been introduced above (Section 2.2.1) as a bridge between the Service Domain that deals with an abstraction of network resources and the Signaling domain that is oriented to control calls/connections.

The session models the coordination of the different calls and connections involved in a single service (and therefore different BCSMs and BCUSMs), as perceived locally in the relevant B-SSF. The IN Switching Manager in the B-SSF handles the requests coming from the BCSMs/BCUSMs and correlates them in the context of a session, sending the appropriate messages to the B-SCF. In the other direction, the IN Switching Manager receives the commands coming from the B-SCF and takes the appropriate actions on the BCSMs or BCUSMs. The IN-SSM is the tool needed to represent the calls, the connections, the parties involved in a session and their relationships.

The concept of session appears very general and powerful to coordinate different connections. Such a concept is mandatory when inter-working with a B-ISDN network implementing Signaling Capability Set 1 which allows only point-to-point single connection calls. When more advanced signaling features will be added to the B-ISDN (e.g. point-to-multipoint, multi-connection), the session concept may slightly reduce its role, as more complex call topologies will be handled directly at the B-ISDN level and therefore in the B-CCF. However, the capability of coordinating different connections belonging to one or more calls will always be a powerful support for the provisioning of advanced services that require to be driven by the network exploiting IN Service Features. Another very promising aspect is that the session concept and the IN Switching Manager functionality allow a smoother transition towards future signaling capability sets. Using the modeling capabilities offered by the IN-SSM, it should be possible to describe a set of operations on the interface between the SCF and the B-SSF as independent as possible from the capabilities offered by the underlying signaling network. The IN Switching Manager processing can map these operations into the corresponding operations on BCSMs over the B-CCF/B-SSF (conceptual) interface and on BCUSMs over the B-CUSF/B-SSF (conceptual) interface. The SSM session views in the B-SSF are 'local' to the different local exchanges or SSPs, using the IN terminology. Another important aspect concerns the 'service independence'. According to the IN principles, the session view should be service independent and it should allow the B-SCF to provide the suitable configurations for a large set of services. The service specific aspects are handled by the service logic in the B-SCF.

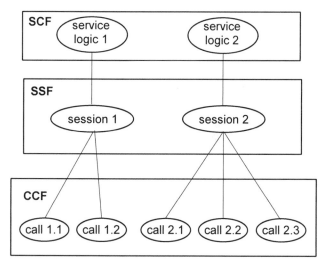

Objects (instances) shown as ovals and denoted in lowercase.

**Figure 2.2.14**
Example of object instances

Assuming that the B-SSF, and in particular the IN Switching Manager, has the role of providing a very comprehensive session view, a suitable model for the IN-SSM has to be developed. For the IN-SSM design an object-oriented modeling has been chosen, where the different parties involved in a session and the user plane/control plane connections are described. This new IN-SSM is powerful enough to represent both 'abstract' information like the ownership of a session, and more 'concrete' elements like the set of basic B-ISDN calls/connections involved in a session. In addition to these general considerations, the introduction of session allows a smooth transition of the traditional functional decomposition used in the IN and B-ISDN standards into a modern object-oriented specification.

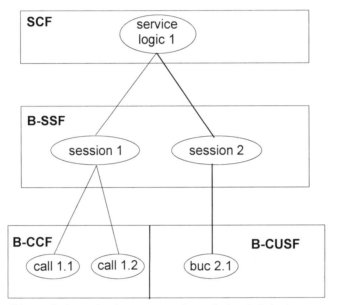

Objects (instances) shown as ovals and denoted in lower case.
'call' = instance of class Call/Bearer Connection
'buc' = instance of class Bearer Unrelated Connection

**Figure 2.2.15**
Example of advanced object instance configuration

Figures 2.2.14 and 2.2.15 give some examples of possible object configurations. In particular, Figure 2.2.14 illustrates how one session may comprise several calls/bearer connections (session 1 and session 2) and Figure 2.2.15 how one service logic may comprise several sessions. In this last example session 1 comprises two call/bearer connections, while session 2 comprises two bearer unrelated connections (session 2 is instanced when an end user invokes a User Service Interaction). In this case the B-SCF functionality should be enhanced in order to comprise the capability of associating two sessions and maintaining the relationship between them.

Another possibility is represented by service combination. This situation is shown in Figure 2.2.16, where the same call is seen by session 1 (call 1.2) and by session 2 (call 2.1). A typical situation could be represented by a user involved at the same time and within the same call in a B-VC service and B-VPN service.

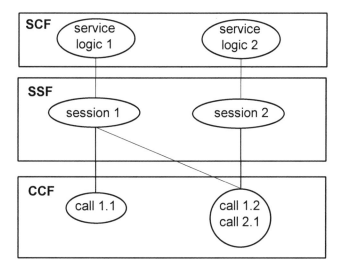

**Figure 2.2.16**
Example of object instances for service combination

## 2.2.5.2 The IN Switching State Model

As introduced in Chapter 1.2, the IN Switching State Model (IN-SSM) of CS-2 is in charge of offering to the SCP an abstraction of the switching and transmission resources allocated in the network for the provision of an IN service.

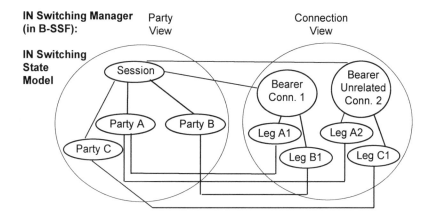

**Figure 2.2.17**
Views realized by Switching State Model

In the integration of IN with B-ISDN, the IN-SSM is assuming an even more central role because it provides a model for the complex call/connection configurations which may appear in B-ISDN. Moreover, the IN-SSM is also able to offer the B-SCF a view of the bearer unrelated connections which are used to transport the information for User-Service Interaction.

Conceptually, the IN-SSM for integrated IN/B-ISDN can be understood as the view of a session and of its associated parties/connections/legs maintained by the IN Switching Manager. Figure 2.2.17 gives an example of an instance of IN-SSM, representing a session with two connections: Bearer Connection 1 between Party A and B and Bearer unrelated Connection 2 between Party A and C. Typically, the Bearer unrelated Connection is used to transport User Service Interaction information, and therefore Party C usually represents the entity which controls the service, i.e. the SCP. As it was mentioned above, the SSM represents two different (but closely related) abstract views, the Party View and the Connection View, which are also indicated in Figure 2.2.17.

In terms of an object-oriented software structure, the bubbles shown in Figure 2.2.17 correspond to *object* instances. Each object encapsulates some amount of local data (attributes) and some functionality (operations). An abstract description of the different kinds of objects (like party, session) is called a *class*. In this book, the notation of OMT (Object Modeling Technique) [9] is used to specify classes (see also Appendix B). Figure 2.2.18, defined below, gives the so-called *class diagram* for the IN-SSM which shows the classes of IN-SSM objects, their attributes and their mutual relationships (which are called *associations* in OMT).

A *Session* object is the representation of a complex call configuration, as it is seen by an IN Service.

A *Party* object represents either an end user or a network component (e.g. the SCP), called a virtual party. The attribute *Is_Virtual* is used to distinguish between real parties and network components (virtual parties). This allows modeling of SCP-initiated actions, for example connection establishment, connection transfer or connection release. The party is also characterized by its *Number* attribute. Between a session and its parties two kinds of associations can be defined: *joins* and *owns*. All the parties involved in the session are said to have joined the session. Only one of these parties is the session owner, so at least one party must join a session. During the life of a session, new parties can be added to or joined parties can be removed from a session.

A *Connection* object represents the abstraction of the user plane or control plane connections that can be established between the parties of a session. In fact, the connection object specializes respectively as a *Bearer connection* or a *Bearer unrelated connection*. The bearer connection objects represent B-ISDN connections. Within B-ISDN Signaling CS-1, there is a one-to-one relationship between call and bearer connection. Therefore, a bearer connection object can also represent a basic CS-1 call. A bearer unrelated connection object represents the control resources needed to perform the dialog between a party and the Service Logic in the B-SCP. This dialog is required to support the User Service Interaction. All connections are characterized by their state, described by the *Status* attribute, but only bearer connections have a *Forward peak bandwidth* and a *Backward peak bandwidth*. Between parties and connections the *ownership* association can be defined so that any connection is owned by one party which is involved in the connection (the *Connection Owner*). The *comprises* association is defined between the session and its connection.

A *Leg* object represents the communication path to a party which is connected to other parties by a connection. Each connection is potentially composed of several legs. The *Bearer leg* and the *Bearer unrelated leg* classes are specializations of the Leg class. Objects of these two classes are both characterized by the attributes *Status* and *C_Plane_Direction,* inherited from the Leg class. But only the bearer leg is characterized by the attribute *U_Plane_direction.* Between parties and legs the association *is connected by* can be defined. A party is connected by legs to all connections in which it is involved. Moreover each leg is in an association with the connection it belongs to. The association *links* is defined between the specialized classes such that a bearer unrelated leg links a bearer unrelated connection and a bearer leg links a bearer connection.

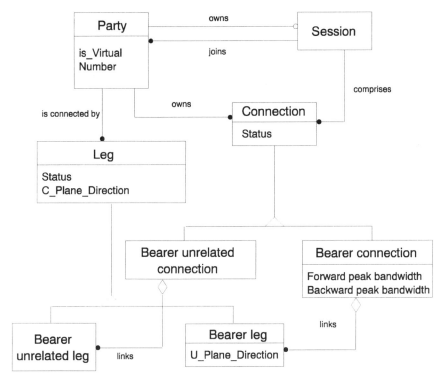

**Figure 2.2.18**
Class diagram for the IN-SSM

Each connection is associated with at least one leg. This relation can be modeled as an aggregation association in OMT. In particular, the multiplicity of the aggregation between a connection and its legs determines the topology type of the connection. If a bearer connection contains exactly two bearer legs, it is a point-to-point connection in the sense of B-ISDN Signaling Capability Set 1. The

IN-SSM model, however, is general enough to handle point-to-multipoint connections as well. A difference between bearer connection and bearer unrelated connection is that the first can be linked to many bearer legs (for point-to-multipoint connections), while the latter is inherently point-to-point: one leg for the communication between Party and B-SSP and the other leg for connecting the B-SCP to the B-SSP.

The attributes of the objects are described in more detail in Table 2.2.1. The attribute *Is_Virtual* indicates whether the respective party object is representing a network element not connected to user plane channels (i.e. the SCP) or a true party associated with an end system (i.e. the Customer Premises Equipment and the B-IP). The attribute *Number* indicates the number used by the B-CCF (or B-CUSF) to identify a Party. It is an E.164 number. This attribute is relevant only if the attribute Is_Virtual is equal to false. The *Status* attributes of connection and leg represent the state of the respective object with respect to call processing. The Bearer connection and Bearer unrelated connection objects inherit the attribute Status from the Connection class while the Bearer leg and Bearer unrelated leg objects inherit the attribute Status from the Leg class.

**Table 2.2.1**
Objects and attributes

| Class name | Attribute | Possible values of attribute |
|---|---|---|
| Session | | |
| Party | Is_Virtual | False |
| | | True |
| | Number | E.164 address (incl. IN Number) |
| | | Point Code |
| Connection | Status | Being setup |
| | | Setup |
| | | Being released |
| Bearer connection | Forward peak bandwidth | # cells/s |
| | Backward peak bandwidth | # cells/s |
| Bearer unrelated connection | | |
| Leg | Status | Pending |
| | | Destined |
| | | Joined |
| | | Abandoned |
| | | Refused |
| | C_Plane_ Direction | Incoming |
| | | Outgoing |
| Bearer leg | U_Plane_ Direction | Source |
| | | Sink |
| | | Bi-Directional |
| Bearer unrelated leg | | |

The state diagrams in Figures 2.2.19 and 2.2.20 show the possible transitions between the values of the Status attribute for the connection and leg object. It should be noted that the state model is not the result of the one-to-one mapping of the corresponding B-CCF states, but depends on the reports of IN-SSM changes requested by the service logic.

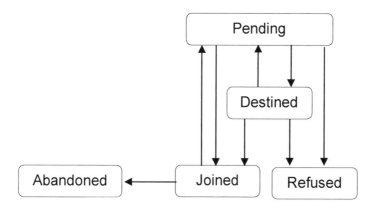

**Figure 2.2.19**
State Diagram for Legs

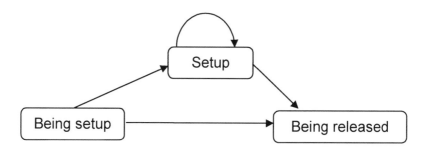

**Figure 2.2.20**
State Diagram for Connections

There is a relationship between the status attributes of a connection and its corresponding legs:
- the status of the connection is 'Being setup', as long as the status attributes of all corresponding legs are not 'Joined';
- if and only if all corresponding legs have status 'Joined', the connection will have status 'Setup'.

The leg attribute *C_Plane_Direction* is used to indicate the direction of the signaling relationship. The attribute value is 'Incoming' for a leg connected to a calling party and 'Outgoing' for a leg connected to a called party. Please note that these values are used from the point of view of the node in which the B-SSF resides. In standards for UNI signaling, an opposite use of these values is common, since there the point of view of the end system is preferred.

Two additional attributes characterize the bearer connection object: the *Forward peak bandwidth* attribute (it indicates the allocated peak bandwidth in the direction from the connection owner to the other party) and the *Backward peak bandwidth* attribute that indicates the allocated peak bandwidth in the direction towards the connection owner. In this way the B-SCF is aware of the bandwidth assigned to the user-initiated bearer connections (while for SCP-initiated bearer connections, the B-SCF explicitly indicate the bandwidth values to be used in the establishment phase).

For user-initiated calls, the connection owner is the calling party: this means the forward direction is from the incoming to the outgoing leg, while the backward is the direction from the outgoing to the incoming leg. In the case of SCP-initiated calls (where two outgoing legs are present) the forward direction is decided by the Service Logic when the connection owner role is assigned to one of the two involved parties.

The flow of user plane information can be indicated by the attribute *U_Plane_Direction* of a bearer leg, which admits values of 'Source', 'Sink', and 'Bi-Directional'. For example, a basic SCS1 call is represented by a bearer connection having two bearer legs with the attribute *U_Plane_Direction* set to 'Bi-Directional'.

According to the principles of object-oriented modeling, the identification of objects does not rely on normal attributes but on implicit identifiers. For the following chapters it is assumed that an object is uniquely identified by a pair (ObjectClass, ObjectID), where ObjectClass is an indicator for the class the object belongs to (Session, Party, Connection, Leg) and where ObjectID is an object instance identifier. The scope of an ObjectID is restricted to objects of the same class within a given session instance, so the same ObjectID may be used in the same session instance for objects of different classes. The subclasses use the same identifier of the class they refer to; for example, Bearer Leg and Bearer Unrelated Leg both use the Leg ID.

The IN-SSM view, including associations between objects, is communicated from B-SSP to B-SCP (or vice versa) through information elements in the Information Flows.

The *session ownership* association in the SSM class is excluded from more detailed specification here. This means that the session ownership association is not represented within both the B-SSF and the B-SCF, and the Information Flows do not transport parameters related to session ownership. This corresponds to the design decisions taken in the project INSIGNIA. Nevertheless, an explicit treatment of session ownership may be useful when for instance advanced charging concepts are considered. This is the reason why the association is present in the IN-SSM model.

The *connection ownership* association in the SSM class diagram is visible both to the B-SCF and to the B-SSF for two reasons:

- To clearly identify the forward and backward direction of user and SCP-initiated calls (the direction is used in the C_Plane Leg Direction attribute and in the peak bandwidths bearer connection attributes);

- To give the connection owner some particular rights: for example, when introducing the bandwidth modification capability, only the connection owner is allowed to start a modification request procedure.

The IN-SSM model is used not only for directly acting on the objects (creation, deletion) but also for handling the monitoring of the status of objects according to the evolution of the B-CCF plane.

## 2.2.5.3 Actions on the IN Switching State Model

The B-SCF/B-SSF dialog refers to IN-SSM objects and states. This means that the operations requested by the B-SCF act on the IN-SSM objects and not directly on the BCSM and BCUSM. It is the B-SSF who is in charge of translating the commands received from the B-SCF into the actions on the appropriate Broadband BCSMs and BCUSMs (as in IN CS-1 and CS-2).

To address operations on the IN-SSM, the B-SCF has to maintain its own view of the IN-SSM. Therefore, a major point for the communication between B-SCF and B-SSF is that the IN-SSM state in the B-SSF has to be aligned with the IN-SSM state seen by the B-SCF. For each event changing the IN-SSM state, a communication flow is needed which provides the information to keep the SSM state aligned in the two network elements.

Two cases are possible:

- The IN-SSM state changes according to an event notified by the B-CCF. This means that in the corresponding Information Flow from B-SSF to B-SCF the new IN-SSM state is notified;
- The IN-SSM state changes according to an operation invoked by the Service Logic Program in the B-SCP. This means that in the corresponding Information Flow from B-SCF to B-SSF the new IN-SSM state change is driven by appropriate Information Elements.

To ease the definition of the information flows, it is necessary to identify a limited set of transitions from one IN-SSM state to another. There are several options how such a set of transitions can be approached. A simple approach would be to provide separate transitions for each possible change (like the creation of an object instance or the update of an attribute). In the following, another approach is used which tries to identify 'larger' transitions which couple together several atomic changes of the IN-SSM such that they are relevant from the call processing point of view. This approach leads to a quite reasonable level of granularity in B-SCF/B-SSF dialog, as can be seen in the next chapter.

Applying this approach, a small number of relevant transitions can be found. IN-SSM state transitions caused by actions requested by the B-SCF are in correspondence with B-SCF invoked operations, while IN-SSM state transitions due to B-CCF events are reported via B-SSF invoked operations.

In Table 2.2.2 there is the set of 'creation' transitions and corresponding operations, while in Table 2.2.3 there is the set of 'deletion' transitions and corresponding B-SCF invoked operations. The terms creation (deletion) mean that, whatever the starting IN-SSM state is, the creation (deletion) transition adds (deletes) a fixed set of objects and relationships, reaching the target SSM state.

**Table 2.2.2**
'Creation' transitions of SSM states

| Transition and corresponding operation | Created objects and relationships |
|---|---|
| C1<br>Join party to session and link leg to bearer | New party X<br>New leg Y 'Pending'<br>leg Y is linked to an existing bearer connection in status 'Being setup' |
| C2<br>Join party and bearer to session | New party X1<br>New leg Y1 'Pending'<br>New bearer connection Z 'Being setup'<br>New leg Y2 (related to an existing party)<br>Legs Y1 and Y2 linked to bearer connection Z |
| C3<br>Add parties and bearer to session | New parties X1 and X2<br>New legs Y1 and Y2 'Pending'<br>New bearer connection Z 'Being setup'<br>Legs Y1 and Y2 linked to bearer connection Z |
| C4<br>Add bearer to session | New legs Y1 and Y2 (related to existing parties X1 and X2) 'Pending'<br>New bearer connection Z 'Being setup'<br>Legs Y1 and Y2 linked to bearer connection Z |

**Table 2.2.3**
'Deletion' transitions of SSM states

| Transition and corresponding operation | Deleted objects and relationships |
|---|---|
| D1<br>Release session | All the objects and related relationships |
| D2<br>Drop party | Party X1<br>All legs Y1,...,Yn connected to X1<br>All connections Z1,...Z2 that remain linked to only one leg<br>All legs that are no more linked to a connection<br>All parties that are no more connected to legs<br>except parties with attribute Is_Virtual equal to True |
| D3<br>Release connection | Connection Z<br>All legs linked to Z<br>All parties that are no more connected to legs except parties with attribute Is_Virtual equal to True |

Three aspects are relevant:

- In every SSM state an instance of a party can exist if and only if it is connected to at least one leg; the only exception is a Party with Attribute

Is_Virtual set equal to True;
- The identified seven transitions are usable for a large variety of services;
- The name of the transition will be used as the name for the Information Flow in the direction B-SCF/B-SSF.

About the last point, the illustrated model implies that if the transitions from an SSM state are driven by B-SCF, the B-SSF creates/deletes objects according to the parameters of the appropriate Information Flow which contain the identifiers of the objects. In the case of creation transitions, this implies that the identifiers of the new objects can be assigned either by B-SSF or B-SCF.

Uniqueness is guaranteed by the following considerations:
- The Object IDs scope is limited by the Session Id, that is the same Object ID can be assigned to different sessions;
- When the first trigger is detected, the B-SSF generates the IDs and notifies B-SCF in the Service Request information flow; for the remaining time of the session, only B-SCF creates new objects and assigns IDs.

Regarding the notification of SSM state changes, a request-report and report mechanism, is used, like in IN CS-1 and CS-2. The main difference is that the report is related not to BCSM events but to changes of objects' attribute values. For more details the reader is referred to Chapter 2.3.

## 2.2.6   Relationship of the Enhanced IN-SSM with Broadband CCF and SCF

In the previous parts the particular role of the Session control in the IN/B-ISDN architecture was emphasized as an additional layer between service control domain and call/connection control domain. In this chapter some considerations about the 'vertical' relationship between layers are given.

### *2.2.6.1 Encapsulation of CCF by the IN Switching Manager*

As already mentioned, the actions performed on the IN-SSM by the Switching Manager are triggered on one side by the B-SCF and on the other side by both the B-CCF and the B-CUSF. The protocol defined in Chapter 2.3 is used to exchange information between the service logic (in the Service Control Domain) and the Session Control Domain. It is an innovative aspect of the proposed architecture that the Service Control Function and in particular the Service Logic do not have a direct view of the states of the Signaling Domain state machines (DPs of BCSM and BCUSM). However, since the actual B-CCF is defined in terms of BCSM and BCUSM some translation is required between the abstract and IN-SSM related view and the lower layer views of the B-CCF and B-CUSF. The IN Switching Manager is in charge of this translation. It is remarkable that in the proposed model the handling of call related events is much more homogeneous for the B-SCF point of view with respect to what happens in IN CS-2, where different interfaces are present between SCF and SSF on one side and between SCF and CUSF on the other side.

The dialog between B-SSF and B-CCF is realized via a dialog between the IN Switching Manager and the Basic Call Manager. This dialog allows each functional entity to exactly determine which actions have to be performed by the manager on the state machine to react to events monitored in the other functional entity. Figure 2.2.21 gives a number of examples for the translation which takes place between Basic Call Manager and B-SCF. Several examples of messages between B-SSF (IN Switching Manager) and B-SCF are shown in the upper part together with the corresponding semantics on the level of the Basic Call Manager. For a complete description of the protocol see Chapter 2.3.

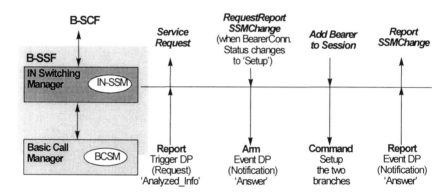

**Figure 2.2.21**
Mapping between call events and session state events

As it was already mentioned in Chapter 2.1, the traditional IN Information Flows appear here again, but are limited to the SSF/CCF interface which is internal to the B-SSF. So the traditional INAP protocol is not required for these Information Flows as long as pure Broadband services are considered. However, it is useful to maintain the traditional INAP structure for the internal interface within the B-SSF, since this enables an easy construction of a hybrid SSF which is able to handle the traditional INAP protocol as well as the enhanced B-INAP.

## 2.2.6.2 Impact of Enhanced SSM on B-SCF

The previous sections have described the rationale for adopting a session concept based on B-ISDN Capability Set 1 basic calls. As a consequence of this decision, it is the B-SSF which controls the network view of an IN service (this is the session control domain). The events of different BCSMs representing different calls are handled and coordinated by the SSF Switching State Model and combined into more abstract messages that are sent to the SCF.

From the view point of the SCF this has a significant impact on the design of the Service Logic Programs. In effect, the Service Logic Programs refer to a global view of the topology of a session. This is an abstracted view mirroring the information which is provided in the IN-SSM in the SSF. The execution of a service in the SCF consists mainly in creating and modifying an appropriate

SSM. Figure 2.2.22 illustrates the effect of the SSM modeling on the B-SCF side.

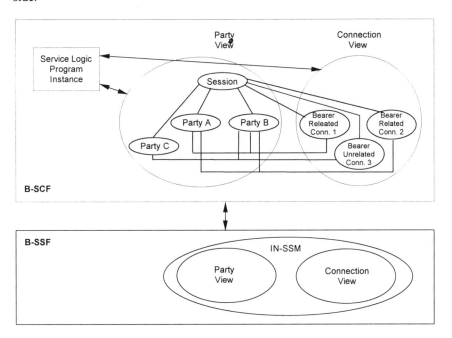

**Figure 2.2.22**
Session models in B-SSF and B-SCF

The session concept keeps the SCF in its original service handling role, leaving control of bearer related and bearer unrelated connection entities to the SSF. Thus it allows for a control relation to be set up between IN and an abstract view of the B-ISDN network provided by the SSF, and minimizes modifications to the already existing SCF Finite State Machine and the SCF/SSF interface. In this respect, the B-SCF provides both service handling and connection services (connection based resources control), thus extending IN visibility to service components and also allowing for easier interaction between IN services, supplementary services and service dealing with connection types.

The B-SCF is able to negotiate the establishment and release of all calls and connections necessary to implement a particular service and also to request any modifications on calls or connections during provisioning of the service. For example, during a Video Conference session, the conference coordinator may wish to add a further user or bearer or change the bit rate, in which case, the B-SCF, through appropriate SLPs, will have to provide terminal profile checks and call barring data, before requesting the B-SSP to actually establish the new connection or modify the existing bearer.

# References

[1]  ITU-T Recommendation Q.1224, *Distributed functional plane for Intelligent Network CS-2*, 1997

[2]  ITU-T Recommendation Q.1218, *Interface Recommendation for Intelligent Network CS-1*, 1995

[3]  ITU-T Recommendation Q.2931, *Digital Subscriber Signaling System No. 2 (DSS2) - User-Network Interface (UNI) layer 3 specification for basic call/connection control*, 1995

[4]  ITU-T Recommendation Q.2761, *Functional description of the B-ISDN user part (B-ISUP) of Signaling System No. 7*, 1995

[5]  ITU-T Recommendation Q.2762, *General Functions of messages of the B-ISDN user part (B-ISUP) of Signaling System No. 7*, 1995

[6]  ITU-T Recommendation Q.2763, *Signaling System No. 7 B-ISDN user part (B-ISUP) - Formats and codes*, 1995

[7]  ITU-T Recommendation Q.2764, *Signaling System No. 7 B-ISDN user part (B-ISUP) - Basic call procedures*, 1995

[8]  ITU-T Recommendation Q.2932.1, *Digital Subscriber Signaling System No. 2 - Generic functional protocol: Core functions*, 1996

[9]  Rumbaugh J, Blaha M, Premerlani W, Eddy F and Lorensen W, *Object-Oriented Modeling and Design*, Prentice Hall, 1991

# Chapter 2.3

# ADVANCED PROTOCOLS FOR MULTIMEDIA SERVICES

The main objective of this chapter is to provide an overview of the control plane of the Intelligent Broadband Network. The description of the physical architecture and of the signaling protocols complements the description of the functional architecture given in Chapter 2.2.

The reference physical network architecture for the Intelligent Broadband Network, which is shown in Section 2.3.1, allows the identification of the physical interfaces between the network elements. Both the control plane and the user plane protocols must be defined to run on these interfaces. In this chapter emphasis is put onto the control plane protocol stacks which are essentially application-independent. It is worth noting that ATM will be used over all the physical interfaces of the Intelligent Broadband Network, and that the lower protocol layers (up to the ATM layer) are common to the control plane and to the user plane protocol stack.

The following chapters describe the protocols which run over the different interfaces, in order to highlight the enhancements that are required according to the Intelligent Broadband Network approach. In particular, Section 2.3.2 deals with the basic B-ISDN protocols showing the control procedures which have been specifically introduced or enriched in the integrated IN/B-ISDN context: the SCP-initiated call and the Bandwidth Modification for SCP-initiated calls. Section 2.3.3 describes the IN application protocols (B-INAP) which run over the B-SCP - B-SSP interface and over the B-SCP - B-IP interface. Finally, some aspects related to the outstanding feature of User-Service Interaction are discussed in Section 2.3.4.

## 2.3.1 ATM Based Intelligent Network Protocol Architecture

### 2.3.1.1 Reference Physical Network Architecture

In this section a reference physical network configuration is provided, in order to identify the reference points (RP) between physical entities (PE). The reference points represent the interfaces where the control plane protocol stack needs to be defined. The first step is to define the mapping of functional entities into physical entities, as shown in Figure 2.3.1. A reference physical configuration for the Intelligent Broadband Network is depicted.

The following physical entities are considered:

*Intelligent Broadband Networks*. Edited by I. S. Venieris, H. Hussmann
© 1998 John Wiley & Sons Ltd.

- TE (Terminal Equipment): This is the user terminal, i.e. a Personal Computer or workstation equipped with an ATM board and located at the user's premises;
- B-SSP (Broadband Service Switching Point): This is the public ATM switch with IN capability. The B-SSP contains the functions of B-SSF, B-CCF and B-CUSF;
- B-IP (Broadband Intelligent Peripheral): This is a device owned by the network operator or service provider, needed to allow user plane interaction between the IN entities and the end users. The B-IP contains the functions of the B-SRF;
- B-SCP (Broadband Service Control Point): This is the IN node that contains the logic for the control of IN services. The B-SCP contains the functions of the B-SCF (and the corresponding SDF).

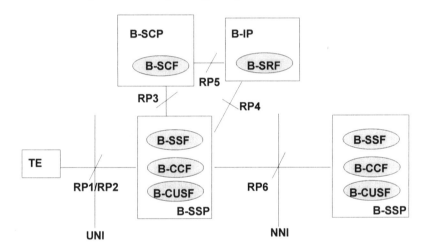

**Figure 2.3.1**
Reference physical network configuration

The following reference points are identified:

- RP1/2 (TE - B-SSP): This is the user Network Interface (UNI) between terminal equipment and the local exchange. It is described in Subsection 2.3.1.2 below. Two reference points are merged because it is assumed that either no access network is used or that the used access network provides transparency for the UNI signaling protocols;
- RP3 (B-SSP - B-SCP): This is the IN Interface between the IN switch and the Service Control point. It is described in Subsection 2.3.1.4 below;
- RP4 (B-SSP - B-IP): This interface is a UNI as well and therefore described in Subsection 2.3.1.2;
- RP5 (B-SCP - B-IP): This is the IN interface between the Intelligent Peripheral and the Service Control Point. It is described in Subsection 2.3.1.4 below;

- RP6 (B-SSP - B-SSP): This is the Network Node Interface (NNI) between two switches. It is described in Subsection 2.3.1.3 below.

## 2.3.1.2 Protocol Stack at the UNI

The control plane protocol stack at the User Network Interface is shown in Figure 2.3.2. Support is provided for basic call/connection control (Q.2931 [1]), for bandwidth negotiation and modification (Q.2962 [2], Q.2963.1 [3]) and for the Generic Functional Protocol in the COBI - Connection Oriented Bearer Independent mode (Q.2932 [4]). This protocol stack is used at the RP1/2 between the Terminal Equipment and the B-SSP, exploiting all the above functionality. It is also used at the RP4 between the B-IP and the B-SSP. This is necessary to handle user-plane connections for the IN services requiring in-band interaction with the service logic. In this case only basic call/connection control functionality is needed.

The Signaling ATM Adaptation Layer (S-AAL) is composed of several layers as described in [5]. Only the higher layer, the Service Specific Control Function (SSCF), is specific for the UNI.

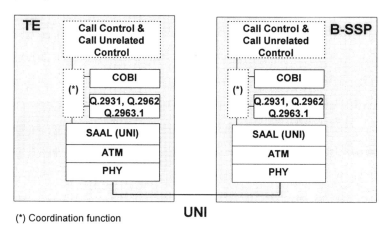

**Figure 2.3.2**
Protocol stack at UNI

## 2.3.1.3 Protocol Stack at the NNI

The control plane protocol stack at the Network Node Interface is shown in Figure 2.3.3. Support is provided for link-by-link call/connection control (Q.2761.4 [6-9]) and for bandwidth negotiation and modification (Q.2725.1 [10], Q.2725.2 [11]). This protocol stack is used at the RP6 between two B-SSPs.

The Signaling ATM Adaptation Layer (S-AAL) for the NNI is composed of the same layers as for the UNI. The SSCF is specific for the NNI. The Message Transfer Part (MTP-3b) is derived from the narrowband MTP-3 for the SS7, with only minor changes. As far as the RP6 is concerned, the MTP-3b can be drastically simplified, excluding the routing functionality. In this case two adjacent B-SSPs will exchange signaling messages over a dedicated link and no forwarding of MTP-3 messages is allowed.

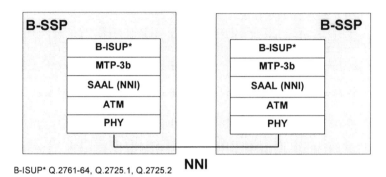

B-ISUP* Q.2761-64, Q.2725.1, Q.2725.2  **NNI**

**Figure 2.3.3**
Protocol stack at NNI

## 2.3.1.4 Protocol Stack at the IN interfaces

The control plane protocol stack at the IN interfaces is shown in Figure 2.3.4. It provides the transport of the IN application protocol (B-INAP). This protocol stack is used at the RP3 between the B-SSP and the B-SCP and at the RP5 between the B-IP and the B-SCP. The lower layers up to MTP-3b are the same as in the NNI protocol stack. The routing functionality in the MTP-3b must be reintroduced in this context to support remote communications between B-SSPs and B-SCPs. The SCCP and the TCAP layer have no specific characteristics with respect to the ones in the narrowband protocol stack.

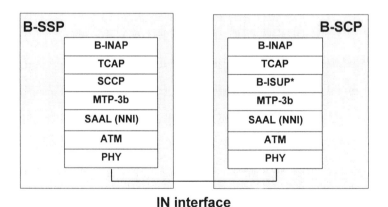

**IN interface**

**Figure 2.3.4**
Protocol stack at the B-SSP - B-SCP interface

## 2.3.2 Advanced B-ISDN Signaling for Integration with IN

The new characteristics of the functional model for Intelligent Broadband Networks as described in Chapter 2.2 impose new requirements to the signaling protocols. In particular the new possibility given to the Service Logic to order the establishment of additional connections has to be considered here. In this chapter an analysis of the enrichment of the signaling capabilities with respect to the standardized protocols (described in Chapter 1.3) is performed.

As a first item, the SCP-initiated calls are examined in Subsection 2.3.2.1 from the point of view of the Call Control Function. In particular in the B-SSP different alternatives are possible which directly reflect on signaling protocols for handling the basic call.

Additional signaling capabilities can be affected by the introduction of SCP-initiated calls: as a significant example, Subsection 2.3.2.2 describes how the bandwidth modification signaling capability is altered when applied to SCP-initiated calls instead of the user-initiated calls.

### 2.3.2.1 B-ISDN Control Procedures for SCP-initiated Calls

The SCP-initiated call is defined as a particular call that is on behalf of an instruction given by IN. It is a special case of a third-party call, since it involves the IN control entity (SCP) as well as normal parties. It is important to underline that IN service logic is only aware of the concept of connection and not of the call, as indicated in Subsection 2.2.4.2 and Section 2.2.5. This means that IN orders the establishment of the new connection and at the signaling level the SSP could decide to map this operation in the set up of a new call or in the adding of a new connection to an existing call.

Considering that the analysis is limited to mono-connection calls, in the following there will be a one-to-one mapping between call and connection but this is not an intrinsic limitation of the model. In Chapter 3.3 some considerations will be given on how to extend the model to handle multi-connection calls.

In the standardized signaling protocols the calls/connections are always established by a calling end system, therefore the Call Control Function inside the node is used to receive a setup indication by a remote signaling entity.

In the case of SCP-initiated calls, a request for establishing a new connection is transmitted to the B-SSP node by means of a B-INAP operation. The node translates this request from IN into the signaling level. Different alternatives can be envisaged.

#### Alternative 1: third party call setup

- The node selects one of the involved parties and elects it as 'pseudo calling party';
- A new signaling procedure (belonging to call control and not to connection control) is specified. This procedure allows communication of the information (e.g. address of called party, traffic characteristics of the

connection, quality of service...) to the 'pseudo-calling party' related to the new connection;

- The 'pseudo calling party' starts a normal user-initiated call establishment procedure, populating the parameters of the SETUP message with the previously received information;
- The call is successfully established when the called party has answered the SETUP message.

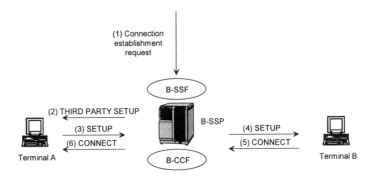

**Figure 2.3.5**
Alternative 1 for SCP-initiated call: third party call setup

*Alternative 2: sequential establishment*
- The node selects one of the involved parties and elects it as 'first called party';
- The node orders the establishment of the 'first half call', that is the part of the call from the node to the 'first called party';
- If the 'first half call' is successfully established, then the procedure for the establishment of the 'second half call' starts;
- The call is successfully established when the 'second called party' has answered the SETUP message.

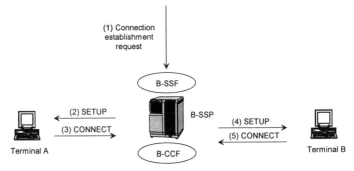

**Figure 2.3.6**
Alternative 2 for SCP-initiated call: sequential establishment

### *Alternative 3: parallel establishment*

- The node orders simultaneously the establishment of the two 'half calls' towards the called parties;
- The call is successfully established when both terminals have answered the SETUP message.

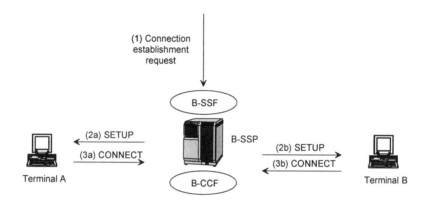

**Figure 2.3.7**
Alternative 3 for SCP-initiated call: parallel establishment

These different possibilities of mapping the command coming from IN into the signaling protocols have advantages and drawbacks.

The main disadvantage of the 'third party call setup' alternative is that a new signaling procedure between the node B-SSP and the terminal must be specified. The advantage is that this procedure can be used for a generic 'third party call setup'; generic means that the third party is not necessarily an IN entity but can be also a signaling entity.

Alternatives 2 and 3 require the introduction of a mirror effect in the B-SSP signaling application process: both parties believe that the party they are connected to is the calling party. Therefore, the inter-working process in the node is different with respect to the corresponding one for a user-initiated call; for example, the CONNECT message from one side must not be forwarded to the other side. The advantage of sequential/parallel establishment is that all the involved network elements (with the only exception of the B-SSP originating the call) are not aware of handling an SCP-initiated call and can use the standard procedures for the basic call handling.

For this reason in this book the first alternative will not be considered further on and in Section 2.5.1 the object model for the B-SSP will refer to alternatives 2 and 3.

It is worth noting that when the SCP-initiated call concept is used together with other advanced signaling capabilities the standard procedures can require enhancements. As an example, in the following chapter the bandwidth modification procedure for SCP-initiated calls is examined.

## 2.3.2.2 Bandwidth Modification Procedures for SCP-initiated Calls

The possibility to modify dynamically the bandwidth of a multimedia channel is a feature offered by the signaling protocols, which support the procedures for the Peak Cell Rate modification of already established point-to-point calls/connections. Such protocols have been specified by the ITU-T recommendations of the B-ISDN Signaling Capability Set 2. In particular, Q.2963.1 [3] is the DSS2 Recommendation series which specifies the Peak Cell Rate modification at the UNI interface, while Recommendation Q.2725.2 [11] specifies extensions to the B-ISUP protocol to support Peak Cell Rate modification at the NNI interface.

These approved standards foresee that only the calling party can request the modification of the bandwidth. The reason for this limitation is to avoid a collision between possible simultaneous requests coming from the two parties.

In the IN/B-ISDN context, the main difference with respect to the standards is that the modification procedure applies also to SCP-initiated calls and not only to user-initiated calls. Considering that the users involved within an SCP-initiated call are both called parties, this means that the standard must be enriched giving both to the calling and to the called parties the possibility to invoke a modification procedure.

This way, a collision of modification requests is possible both for user-initiated calls (calling and called parties simultaneously invoking a modification request) and for SCP-initiated calls (both called parties simultaneously invoking a modification request). In the following, a solution is described which uses IN features in order to solve the problem of the collision of modification requests. In particular, the meaning of connection ownership (given to one of the parties attached to the connection) used in this context differs from the corresponding one adopted in the above mentioned ITU-T signaling specifications, where the owner is always the calling party. Since in the IN-SSM model the Service Logic assigns which party plays the role of the connection owner, the B-SSP can apply the rule for which:

- A modification request coming from the connection owner side is allowed;
- A modification request coming from the non-connection owner side is rejected.

It must be emphasized that only the Application Process of the B-SSP node that originated the SCP-initiated call can apply this rule because the connection ownership concept is not explicitly known by the protocol signaling entities.

As an example, in Figure 2.3.8 the successful modification procedure is shown. The modification is requested by user A which has been designated as the 'connection owner' by the Service Logic residing in the SCP.

It is observable that the forward arrow flips direction when crossing the network: this requires that the Call Control Function in the B-SSP maps the parameters of the incoming MODIFY REQUEST message related to the forward direction in the corresponding backward parameters of the outgoing MODIFY REQUEST and vice versa.

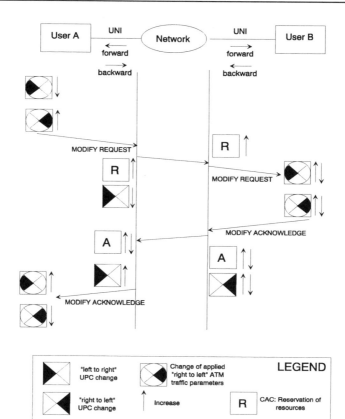

UPC : Usage Parameter Control
CAC : Call Admission Control
**forward** : direction from calling user to called user
**backward** : direction from called user to calling user

**Figure 2.3.8**
Bandwidth modification procedure

The introduction of the bandwidth modification procedure is useful to show that:

- For user-initiated calls no impact is generated on the Intelligent Network side. This means that the introduction of new features in the signaling plane does not necessarily require modifications at the IN level;
- For SCP-initiated call the concept of the connection ownership can be used both by the Service Logic (to assign rights) and by the signaling (to manage permissions);
- It is possible that the Service Logic ignores the bandwidth parameter values related to the connection being established. In this case it is possible to

command the setup of a zero bandwidth connection between the involved terminals (it corresponds to the establishment of a 'control relationship' between the two parties). In a subsequent step, the terminals exploit the modification procedure for the assignment of the bandwidth effectively needed by the application, reporting this change (if this notification has been previously requested) to the Service Logic.

### 2.3.3    Broadband Intelligent Network Application Protocol (B-INAP)

The aim of this chapter is to describe the B-INAP protocols defined on the interfaces between two B-SSPs (RP3 identified in Figure 2.3.1) and between the B-IP and the B-SCP (RP5 identified in Figure 2.3.1). Technically, the B-INAP is a collection of *operations* with related *parameters*, described with the ASN.1 - Abstract Syntax Notation [12]. The B-INAP belongs to the Physical Plane (PP) of the IN conceptual model. For the sake of simplicity, in this section the Information Flows (IFs) and Information Elements (IEs) between functional entities belonging to the IN Distributed Functional Plane (DFP) are described. A one-to-one mapping between Information Flows and B-INAP operations and between Information Elements and B-INAP parameters is assumed. Under this assumption, it is equivalent to refer to the B-SSF - B-SCF relationship on the Distributed Functional Plane or to the B-SSP - B-SCP interface on the Physical Plane.

It is worth noting that according to IN principles, the Information Flows (and the corresponding B-INAP operations) have been defined as service independent as possible, allowing a limited set of information flows to support a large variety of services.

#### 2.3.3.1 B-SSP - B-SCP Interface

This interface (or the corresponding B-SSF - B-SCF relationship) is based on the session model. The basic concepts related to the operations on the session model have been described above in Section 2.2.5. Table 2.3.1 shows the list of the Information Flows (IF) for the B-SSF - B-SCF relationship.

**Table 2.3.1**
B-SSF - B-SCF information flows

|    | IF - Information Flow (Operation)           | Direction              |
|----|---------------------------------------------|------------------------|
| 1  | Service request                             | SSF $\rightarrow$ SCF  |
| 2  | Request report SSM change                   | SCF $\rightarrow$ SSF  |
| 3  | Report SSM change                           | SSF $\rightarrow$ SCF  |
| 4  | Join party to session and link leg to bearer| SCF $\rightarrow$ SSF  |
| 5  | Continue                                     | SCF $\rightarrow$ SSF  |
| 6  | Add bearer to session                       | SCF $\rightarrow$ SSF  |
| 7  | Add parties and bearer to session           | SCF $\rightarrow$ SSF  |
| 8  | Join party and bearer to session            | SCF $\rightarrow$ SSF  |
| 9  | Release Session                             | SCF $\rightarrow$ SSF  |
| 10 | Drop session                                | SCF $\rightarrow$ SSF  |
| 11 | Release Connection                          | SCF $\rightarrow$ SSF  |

| 12 | Drop party | SCF → SSF |
| 13 | SendSTUI | SCF → SSF |
| 14 | RequestReportUTSI | SCF → SSF |
| 15 | ReportUTSI | SSF → SCF |

An instance of a dialog between two B-SSFs for the provision of an IN service is initiated by the B-SSF, by sending a *Service Request* IF. The B-SSF is reacting to a Trigger Detection Point processed in the B-CCF, typically an IN number has been 'dialed' and a call is waiting for instruction. The protocol described here does not yet support the capability to initiate a dialog in the B-SCF → B-SSF direction. But the protocol could be extended easily to support such a capability; for example a *Create Session* IF could be defined.

The B-SCF analyzes the *Service Request* and a Service Logic Program will start up. The Service Logic can now give instructions to the B-SSF by sending one of the information flows which operate on the session objects. For example the *Join party to session and Link leg to bearer* IF will instruct the B-SSF to complete a suspended call providing the destination address (similar to the *Connect* IF in narrowband INAP). The B-SCF can also send *Request report SSM change* information flows. These are needed to request from the B-SSF that it notifies the B-SCF later when specific events have happened, for example that a specific call has/has not been setup. Typically an operation to create an object in the session is associated to a *Request report SSM change* IF to monitor the result of the operation.

During the lifetime of an active dialog, the B-SSF will execute the requests coming from the B-SSF by translating them into operations at the B-CCF level like establish connections or arming DPs in the BCSM. The B-SSF will report the requested notification to the B-SCF with a *Report SSM change* IF.

The *Continue* IF is used to continue the previously suspended call processing without any changes to the SSM object model ('no-operation'). For example, this is necessary when the Service Logic must authorize the call processing to perform a given operation.

The *Release Session* IF is used to terminate the dialog between B-SSF and B-SCF, and to release all the resources controlled by a session. The *Drop Session* IF is used to terminate the relationship between B-SSF and B-SCF without releasing the associated calls/connections. This could be adopted for services like Green Number where the only role of IN is to perform number translation.

The parameters of the Information Flows are defined in such a way that the two views of the session state which are present in B-SCF and B-SSF can be kept consistent (see Chapter 2.2). For this purpose, two kinds of visibility of IN-SSM object associations can be identified. *Explicit* visibility means that there is a data structure in the messages which is a direct representation of the association. *Implicit* visibility means that the information on the association can be derived indirectly from the information elements present in the messages. Only the'link', 'connect' and 'connection ownership' associations are explicitly referred. Table 2.3.2 summarizes this concept.

**Table 2.3.2**
Visibility of associations between objects

| Relationship | visible to SCF | involved information elements (see below) |
|---|---|---|
| Join (session-parties) | Implicit | Party id list |
| Comprise (session-connections) | Implicit | Connection list |
| Link (leg-connection) | Explicit | SSM state |
| Session ownership | Explicit | Session owner id |
| Connection ownership | Explicit | Connection owner id |
| Connect (party- leg) | Explicit | SSM state |

A high level description of the Information Flows is given hereafter. For each Information Flow (IF) the list of mandatory (M) and optional (O) Information Elements is given. Some Information Elements (IEs) are explained in more detail at the end of the Information Flow list. Moreover, at the end of this chapter a small example shows the practical use of the Information Flows.

### List of Information Flows
*(1) Service request (B-SSF → B-SCF)*
This IF is generated by the B-SSF when a trigger is detected at any DP in the BCSM or BCUSM and a new IN-SSM instance is created. It means that an SLP within the B-SCF must be addressed. This IF carries the following Information Elements:

| | |
|---|---|
| Session ID | (M) |
| Service key | (M) |
| Calling party number | (O) |
| Called party number | (M) |
| USI information | (O) |
| SSM state | (M) |

*(2) Request report SSM change (B-SCF → B-SSF)*
This IF is used to request the B-SSF to monitor session related events and consequently SSM state changes, and to instruct the B-SSF to send a notification back to the B-SCF for each detected event. See below for an explanation how the IEs 'Object ID' and 'Object state' are used to specify the state changes. This IF carries the following IEs:

| | |
|---|---|
| Session ID | (M) |
| Object type | (M) |
| Object ID list | (M) |
| Object state list | (M) |

*(3) Report SSM change (B-SSF → B-SCF)*
This IF is generated by the B-SSF to notify the B-SCF that a change in the SSM has occurred. It is sent in response to the Request report SSM change. This IF carries the following IEs:

| | |
|---|---|
| Session ID | (M) |
| Object type | (M) |
| Object ID | (M) |
| Object state | (M) |

*(4) Join party to session and link leg to bearer (B-SCF → B-SSF)*
This IF is used to request the B-SSF to join a new party to an already established session, composed by the calling party with the corresponding leg and a bearer connection, and to link the new leg to the existing bearer connection. This IF carries the following IEs:

| | |
|---|---|
| Session ID | (M) |
| Bearer endpoint | (M) |
| Connection ID | (M) |

*(5) Continue (B-SCF → B-SSF)*
This IF is used to request the B-SSF to proceed after an interruption notified by a state change of the given object. This IF carries the following IEs:

| | |
|---|---|
| Session ID | (M) |
| Object type | (M) |
| Object ID | (M) |

*(6) Add bearer to session (B-SCF → B-SSF)*
This IF is used to request the B-SSF to add a new bearer connection and corresponding legs between existing parties within an already established session. This IF carries the following IEs:

| | |
|---|---|
| Session ID | (M) |
| Bearer endpoint list | (M) |
| Connection ID | (M) |
| Bearer characteristics | (M) |
| Connection owner ID | (M) |

*(7) Add parties and bearer to session (B-SCF → B-SSF)*
This IF is used to request the B-SSF to add a new bearer connection and corresponding legs, within an already established session, between the new parties specified in the IF. This IF carries the same IEs as the *Add bearer to session* IF.

*(8) Join party and bearer to session (B-SCF →to B-SSF)*
This IF is used to request the B-SSF to add a new bearer connection and corresponding legs between (at least) an existing party and (at least) a new party. This IF carries the same IEs as the *Add bearer to session* IF.

*(9) Release Session (B-SCF → B-SSF)*
This IF is used to request the B-SSF to release the whole session. This implies releasing all parties, related legs and bearers and the session itself. This IF carries the following IEs:

| | |
|---|---|
| Session ID | (M) |
| Cause | (O) |

*(10) Drop Session (B-SCF → B-SSF)*
This IF is used to request the B-SSF to release the session (i.e. the logical association between the B-SSF and the B-SCF for a given service instance), without releasing any physical resource. This IF carries the following IE:

| | |
|---|---|
| Session ID | (M) |

*(11) Release Connection (B-SCF → B-SSF)*
This IF is used to request the B-SSF to release the specified bearer connections or bearer-unrelated connections from an existing session. This IF carries the following IEs:

> Session ID                     (M)
> Connection ID list             (M)
> Cause                          (O)

*(12) Drop party (B-SCF → B-SSF)*

This IF is used to request the B-SSF to drop the specified parties, from an existing session. All the objects related to the given party are removed. This IF carries the following IEs:

> Session ID                     (M)
> Party ID list                  (M)

The IFs number 13 through 15 from Table 2.3.1 are related to User-Service Interaction and will therefore be described further below in a separate section devoted to this issue.

### List of Information Elements

The Information Elements that are worth to be commented are listed below. Note that several Information Elements are declared as a list of IE, for example the IE *Object ID list* is a list of *Object ID* IEs.

*Session ID:* This is a session object identifier that allows both B-SCF and B-SSF to uniquely identify the session instance. The session ID is generated by the B-SSF.

*Service key:* This IE is used to address the correct application/SLP within the B-SCF.

*USI Information:* This is a generic data container conveying information from the CPE to SCP and vice versa, see also Section 2.3.4 on User-Service Interaction.

*SSM state:* The SSM state describes a session instance state in terms of existing objects and corresponding relationships. The SSM state is described by a Party description list, a Leg description list, a Bearer description List and a Bearer unrelated connection description list. Although both connection lists are optional, at least one of these lists must be present. As the lists are not nested, some information has to be duplicated for describing relationships. The SSM is a structure containing the following IEs:

> Party description list          (M)
> Leg description list            (M)
> Bearer description list         (O)
> Bearer unrelated
> connection description list     (O)

*Object type:* This IE defines which object class is considered in the monitoring operation. The *Object type* IE can assume one of the following values: *leg* or *connection*.

*Object ID:* This IE supplies the specific object identifiers that have to be monitored for a given Object type.

*Object state:* The *Object state* IE is a structure containing the following IEs:

> Attribute                       (M)
> Attribute value                 (M)
> Monitor mode                    (O)
> Monitoring info requested       (O)

                          Monitoring info reported     (O)

Monitor mode IE is mandatory in the Request Report SSM Change IF and missing in the Report SSM Change. Monitoring info requested IE is present only in Request Report SSM Change IF. Monitoring info reported is only present in Report SSM Change IF.

*Bearer endpoint:* The Bearer endpoint is a structure which combines Party description and Leg description structures in order to describe only the necessary information items. It describes the associations of bearer connection objects with the objects related to a party, and therefore it contains the following IEs:

| | |
|---|---|
| Party ID | (M) |
| Leg ID | (M) |
| U plane direction | (M) |
| C plane direction | (M) |
| Party number | (O) |
| High layer information | (O) |
| Low layer information | (O) |

*Connection ID, Party ID, Leg ID:* These are the identifiers of the object instances within a session. They are used by both B-SSF and B-SCF to uniquely identify the corresponding object. They can be generated either by the B-SSF or by the B-SCF.

*Bearer characteristics:* This is a structure containing the following IEs:

| | |
|---|---|
| Forward peak cell rate | (M) |
| Backward peak cell rate | (M) |
| Bearer class | (M) |
| AAL type | (M) |

### Example for a B-SCF - B-SSF Dialog

An example may be helpful to explain the practical use of the IFs defined above. In the following, a very simple 'second connection service' is presented as an example. In this service the service logic in the B-SCP performs a 'number translation' to connect the calling party to a specific called party, and then the B-SCP instructs the B-SSP to establish an additional bearer connection between these two parties. An interaction diagram is shown in Figure 2.3.9. Concerning the UNI, only the messages on the interface with the Calling Party are shown in the diagram.

A calling user sends a SETUP containing an IN-number to the B-SSP. The IN is triggered during call processing, and the B-SSP invokes a corresponding service at the B-SCP by sending a 'Service Request' message. To keep track of the call, the B-SCP requests for a notification of the setup or release of the bearer connection (Request Report SSM change*)*. Then, the network address is provided within the message 'Join Party To Session And Link Leg To Bearer' to complete the call setup. The B-CCF resumes call processing and notifies the B-SSF when the first call is active. The B-SSP sends a corresponding 'Report SSM Change' message to the B-SCP. Now (or alternatively earlier in parallel to the previous actions) the B-SCP orders the establishment of a second connection between the involved parties, sending a 'Request Report SSM change' and an 'Add bearer to session'. Also the second bearer connection is established successfully and a

'Report SSM Change' is sent to the B-SCP. After some time one of the basic calls is released and this is reported to the SCP which releases the whole session according to the service logic.

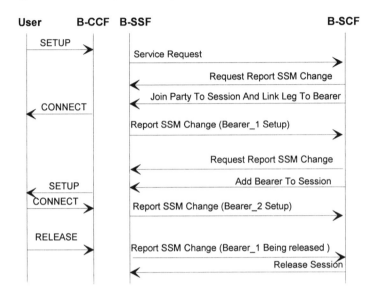

**Figure 2.3.9**
Interaction diagram of a 'Second Connection Service'

## 2.3.3.2 B-IP - B-SCP Interface

The B-IP - B-SCP interface is used to support in-band interaction with a user terminal and allows the exchange of service related information for the services which are provided by the combination of B-SCP and B-IP computing resources. The aspect of distribution of intelligence between B-SCP and B-IP will be discussed in Chapter 2.4 below for concrete services. Here just the basic mechanism for B-IP - B-SCP communication is introduced. Table 2.3.3 shows the list of the Information Flows (IF) on the B-SRF - B-SCF relationship. As above, a high level description of the information flows is then provided. For each Information Flow (IF) the list of mandatory (M) and optional (O) Information Elements (IEs) is given.

**Table 2.3.3**
B-SRF - B-SCF Information Flows

|   | IF - Information Flow (Operation) | Direction |
|---|---|---|
| 1 | Assist Request Instructions | SRF → SCF |
| 2 | Collected User Information | SRF → SSF |
| 3 | Play Announcement | SCF → SRF |
| 4 | Prompt And Collect User Information | SCF → SRF |

### List of Information Flows

*(1) Assist Request Instructions (B-SRF → B-SCF)*

This IF is sent by the B-SRF to the B-SCF, when the B-SRF has received an incoming call from a B-SSF/CCF as a result of the B-SCF sending a *Join Party to session and Link Leg to Bearer* IF, *Add Bearer To Session* IF or *Join Party And Bearer To Session* IF to the B-SSF. An assisting B-SRF sends this information flow to the B-SCF in order to obtain user interaction instructions. This IF carries the following IEs:

| | |
|---|---|
| Correlation Id | (M) |

*(2) Collected User Information (SRF → SCF)*

This IF is sent as the response to the IF *Prompt and Collect User Information*. It contains the information collected from the user. Note that in the INAP, this IF maps on to the RESULT part of the *Prompt and Collect User Information* operation. The SRF sends this IF to the SCF to provide information collected from the user. This IF carries the following IEs:

| | |
|---|---|
| Received Information | (M) |
| Extensions | (O) |

*(3) Play Announcement (SCF to SRF)*

This IF may be used for in-band interaction with a user. The SCF sends this IF to an SRF to initiate user interaction. This IF carries the following IEs:

| | |
|---|---|
| Information To Send | (M) |
| Disconnection from IP Forbidden | (M) |
| Request Announcement Completed Indication | (M) |
| Extensions | (O) |

*(4) Prompt And Collect User Information (SCF to SRF)*

This information flow is used to interact with the user in order to collect information. The B-SCF sends this information flow to a B-SRF to initiate user interaction for a two-party call segment in an B-SSF. This IF carries the following IEs:

| | |
|---|---|
| Collected Info | (M) |
| Disconnection From IP Forbidden | (M) |
| Information To Send | (O) |
| Extensions | (O) |

### List of Information Elements

The main Information Elements used for the B-SCF - B-SRF interface are listed hereafter with a brief description:

*Correlation Id:* This IE is used by the B-SCF to associate the assisting SSF with the *Service Request* from the initiating SSF. The Correlation Id is supplied by the B-SCF to the B-SSF which in turns forwards it to the B-SRF.

*Extensions:* This IE is used as an optional field which allows service specific data to be exchanged between different functional entities, i.e. B-SCF and B-SRF in this case. As an example, three extension fields for use by the Broadband Video Conference service are the *Conference Information Record* (CIR) IE, *User Record* IE and the *Conference Identifier* (CID) IE. The CIR IE allows the

SCF to receive the CIR information from the B-SRF. The User Record IE allows the B-SCF to receive the User Record information from the B-SRF. A complete description of these IEs is out of the scope of this book.

*Information To Send:* This IE is used to specify what information the B-SRF should send to the end user.

*Disconnection from IP Forbidden:* This IE is used to inform the B-SRF whether it can release the connection to the B-SSF/CCF after the announcement has been completed.

*Request Announcement Completed Indication:* This IE indicates that the SRF should send a *Specialized Resource Report* IF when the announcement is complete.

## 2.3.4    User-Service Interaction

### *2.3.4.1 Overview*

User-Service Interaction is a new capability introduced in IN Capability Set 2. There are two basic variants of this capability:

- *Out Channel Call Related User Interaction (OCCRUI)*: This provides a transparent transfer of service information between a user and service logic within the context of a call.  This transparency means that neither the network nor the access signaling protocols analyze the information conveyed, it is just routed to the SSF from where it is transparently transferred to the SCF. User to Service Information may be conveyed within Basic Call Control messages and within Transport Messages (i.e. during the active and alerting phases of a call);

- *Out Channel Call Unrelated User Interaction (OCCUUI)*: This provides the capability to exchange information between the user and the service which is outside the context of a call (e.g. such as registration of a user location). As with the call related case the information is conveyed transparently by the access and network signaling protocols. Unlike the call related case, the information in the form of either a component or embedded PDU, is routed to the SCF via a CUSF (Call Unrelated Service Function). It should be noted that this mechanism will be further enhanced as part of IN Capability Set 3, to allow for User-Service Information IE to be conveyed in the same manner as the Call Related case. This enhancement has been assumed for use in a Broadband environment and is described below.

To provide this capability, the technique used is to envelope the user-service information within the protocols. This is a very simple concept but an extremely powerful tool as the use of such a mechanism allows new IN services to be introduced without waiting for the supporting UNI and NNI protocols to be upgraded to support new IN service specific features. It also allows greater scope and flexibility for service customization of the user interface. The same advantages can be applied to the Broadband environment. The usage of User-Service Interaction in the design of Broadband services is further explored in Chapter 2.4.

## 2.3.4.2 Extension of B-INAP for User-Service Interaction

The data container or the *USI Information* Information Element (IE) as defined by IN CS-2 Recommendations [13] is a variable length octet string having a maximum size. The contents of this IE are evaluated by the IN service logic and the CPE. Although ITU standards do not specify the contents of the USI information IE, the following is one possible structure for the data container, it is based on the ITU message structuring recommendations, i.e. operation, message length and information elements.

| Service Identifier | Operation Identifier | Operation Length | A series of service specific parameters |
|---|---|---|---|

Each USI information IE can carry one service specific operation, the purpose and format of which can be determined from a unique identifier in the *operation identifier* field of the container. The *service identifiers* field identifies the service for which the operation is applicable. The *operation length* field specifies the length of the operation and the remainder of the data container is then constructed as *a series of service specific parameter* information elements. These information elements are each similarly sub-divided into an identifier, length and contents. The information contained in the USI information IE must be agreed between the Service logic and the user terminal provider.

If only Call Unrelated User-Service Interaction is required (as for example in the trials of the INSIGNIA project), the User-Service Information IE can be conveyed in the COBI messages only. However, if Call Related User-Service Interaction is required then the User-Service Information IE can be conveyed in the COBR messages.

Two operations have been defined in B-INAP to convey the User-Service Information IE; *ReportUTSI* and *SendSTUI*. A further operation (with the name *RequestReportUTSI*) has been defined which is used by the B-SCP to instruct the B-SSP to monitor for User-Service Information IE which are received from the user. The operations and their information elements are listed hereafter, the numbering refers to Table 2.3.1 which gives the full list of B-INAP Information Flows.

### List of Information Flows
*(13) SendSTUI (B-SCP → B-SSP)*
This operation allows the B-SCF to send a STUI element to the CPE. This operation contains the following IEs:

      Session ID            (M)
      Leg ID                (M)
      USI Information     (M)

*(14) RequestReportUTSI (B-SCP → B-SSP)*
This operation is sent by the B-SCF to request the B-SSF to monitor for a UTSI message. This operation contains the following IEs:

      Session ID             (M)

                  Leg ID                (M)

                  Monitor Mode     (M)

*(15) ReportUTSI (B-SSP → B-SCP)*

This operation is sent by the B-SSP when an UTSI IE has been received, if monitoring has been previously requested (*RequestReportUTSI*). This operation contains the same Information Elements as the*Send STUI* operation.

### *Example for B-SCF - B-SSF Dialog with User-Service Interaction*

The following example, as depicted in the interaction diagram of Figure 2.3.10, shows a Session initiated by a COBI-connection realizing the USI feature.

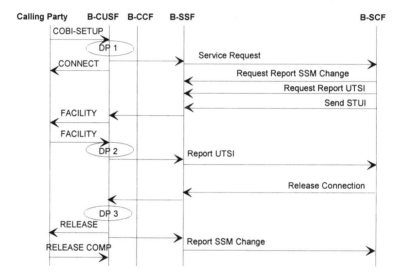

**Figure 2.3.10**
Interaction diagram illustrating the use of User-Service Interaction

    A user sends a COBI-SETUP containing an IN-number to the CUSF. This event passes BCUSM Detection Point 1, where the CUSF requests the SSF to setup a USI relationship between the user and the SCP. For this, the SSF invokes the corresponding service at the SCF by sending a 'Service Request' message. To keep track of the COBI-connection, the SCP requests for a notification of the release of this connection (by the message 'Request Report SSM Change'). This request is transformed by the SSF such that the corresponding BCUSM detection point 3 for this COBI-connection is armed. Then, the SCF enables the user to send USI-information (by the message 'Request Report UTSI'). This again is translated into a BCUSM DP (DP 2) to be armed.

    Finally, the SCF sends USI-information (within the message 'Send STUI') which is forwarded to the corresponding user (within the FACILITY-message).

    At any time during the USI-relationship, the user can send USI-information (via a FACILITY-message) towards the SSP. The SSP forwards this USI-information to the SCF (by a 'Report UTSI'), if the corresponding DP 2 is armed. In the context of the existing session, the SCF could also send other operations, as needed for example to add bearer connections. This is not shown in the example.

The SCF terminates the USI-relationship by sending a 'Release Connection' message. The release of the COBI-connection is detected (DP 3), forwarded to the user (by a RELEASE message) and reported to the SCF (by a 'Report SSM Change' message).

# References

[1]   ITU-T Recommendation Q.2931, *Digital subscriber signaling system No. 2 (DSS2) - User Network interface (UNI) layer 3 specification for basic call/connection control*, 1995

[2]   ITU-T Recommendation Q.2962, *DSS2 - Connection characteristics negotiation during call/connection establishment phase*, 1997

[3]   ITU-T Recommendation Q.2963.1, *DSS2 - Connection modification: Peak cell rate modification by the connection owner*, 1996

[4]   ITU-T Recommendation Q.2932.1, *B-ISDN DSS2 Generic functional protocol: Core functions*, 1996

[5]   ITU-T Recommendation Q.2100, *B-ISDN signalling ATM adaptation layer (SAAL) overview description*, 1994

[6]   ITU-T Recommendation Q.2761, *Functional description of the B-ISDN user part (B-ISUP) of Signalling system No. 7*, 1995

[7]   ITU-T Recommendation Q.2762, *General Functions of messages and signals of the B-ISDN user part (B-ISUP) of Signalling System No. 7*, 1995

[8]   ITU-T Recommendation Q.2763, *Signalling System No. 7 B-ISDN User Part (B-ISUP) – Formats and codes*, 1995

[9]   ITU-T Recommendation Q.2764, *Signalling System No. 7 B-ISDN User Part (B-ISUP) – Basic call procedures*, 1995

[10]  ITU-T Recommendation Q.2725.1, *B-ISDN User part – Support of negotiation during connection setup*, 1996

[11]  ITU-T Recommendation Q.2725.2, *B ISDN User Part – Modification procedures*, 1996

[12]  ITU-T Recommendation X.680, *Information technology – Abstract Syntax Notation One (ASN.1): Specification of basic notation*, 1994

[13]  ITU-T Recommendation Q.1228, *Interface Recommendation for intelligent network CS-2*, 1997

# Chapter 2.4

# INTELLIGENT NETWORK SERVICE SPECIFICATION

In this chapter, the Intelligent Network modeling principles and related protocols, which were introduced in Chapters 2.2 and 2.3, are applied to the specification of the IN services as described in Chapter 1.1 and [1]. Some exemplary service scenarios of each IN service illuminate the interworking of the different Functional Entities. The discussion of the Broadband Virtual Private Network service shows that the new modeling concepts apply well to IN services already available for narrowband networks (see also Chapter 3.4). The novel IN concepts of 'session', 'SCP-initiated calls' and 'User-Service Interaction' serve as powerful tools for the design of an Interactive Multimedia Retrieval and a Broadband Video Conference service.

For the description of the scenarios, interaction diagrams (see Appendix B) are used. Within the textual description of these diagrams, the parameters of the INAP messages which are exchanged during the scenario are provided for a deeper understanding of these messages.

As explained above, the IN-SSM models the call/connection configuration at the CCF level in a more abstract way and presents this view to the SCF. The SCF itself can act on the IN-SSM objects, what is reflected by actions within the CCF. In summary, the actual states of instances of the IN-SSM depend on messages received by the SSF either from the CCF or from the SCF. Therefore, the instances of the SSM are depicted by instance diagrams (see Appendix B) at selected points of time during the scenarios.

The three services are described in different levels of detail. While a complete description of the scenarios for the Broadband Virtual Private Network service is provided, only some interesting parts of the Interactive Multimedia Retrieval scenarios are described in more detail. Finally, the Broadband Video Conference service is discussed only on an abstract level.

## 2.4.1 Broadband Virtual Private Networks

As a general network service, the Broadband Virtual Private Network (B-VPN) service is not related to a particular application, but can be used implicitly by any application which requires point-to-point connections. Therefore, only a small subset of the IN/B-ISDN functionality is needed to realize the B-VPN service, which also is already available for narrowband IN. However, this service

*Intelligent Broadband Networks*. Edited by I. S. Venieris, H. Hussmann
© 1998 John Wiley & Sons Ltd.

provides a good starting point to show the concepts of an integrated Intelligent Broadband Network.

Additionally, independent applications can be provided for the administration of the B-VPN service. These applications make intensive use of the User-Service Interaction (USI) feature (see Section 2.3.4), which is an already standardized, but not widely used IN feature.

After a general description of the B-VPN service, two examples of working B-VPN services are given below. The first example, describing an admitted and successfully established B-VPN call, illuminates the application of the IN/B-ISDN concepts on a simple configuration, consisting only of a point-to-point connection.

The second example, the modification of the VPN routing information by the Follow Me function, shows an application of the USI feature.

### 2.4.1.1 Introduction

The B-VPN service realizes a logical sub-network of a Broadband ISDN which appears to a specific group of users as a private Broadband network. The service is particularly suitable for organizations with geographically wide-spread locations.

Figure 2.4.1 depicts the general situation. VPN customers are represented by the two user groups X and Y, each of which forms a VPN. Locations A2, B1 and C2 are on-net locations for VPN X, B2 and C3 are on-net locations for VPN Y.

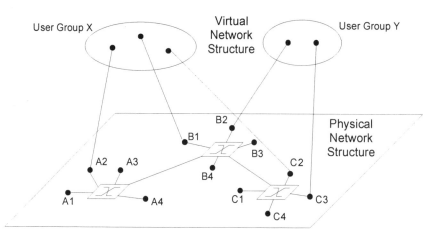

**Figure 2.4.1**
Physical and virtual (logical) networks

Three kinds of VPN calls can be distinguished:
- On-Net to On-Net (or simply On-Net) calls between two users belonging to the same VPN;
- On-Net to Off-Net calls from a user within a VPN towards a user not belonging to the VPN of the calling user;

- Off-Net to On-Net calls towards a user within a VPN from a user not belonging to the VPN of the called user.

IN provides mechanisms to realize a VPN [2]. Some of the VPN functions using the IN features are listed in the following:

- *Private Numbering Plan*: A VPN customer uses a numbering plan which defines private numbers for VPN users and locations in the B-VPN (on-net locations). These private numbers, used within signaling messages from a calling user, trigger the IN service.

- *Call Routing Based upon Number Translation*: Besides a simple static translation between a VPN private number and a public number, more complex algorithms can be used to make the translation dynamically dependent on various parameters. Examples for such algorithms are Time Dependent Routing (TDR) for which the call routing is dependent on Time of Day (ToD) and/or Day of Week (DoW) and the Origin Dependent Routing (ODR) with a routing dependent on Area of Origin (AoO). For company VPNs, mainly TDR is useful; however, ODR within a VPN may also be used in special cases (e.g. if the participants of a client-server type service are organized in a VPN and the servers are geographically distributed).

- *VPN Profile Screening*: Within a VPN, several privileges can be identified, e.g. to use particular information services or to establish calls to privileged numbers (within the VPN). The subscription tables can be used to store the information which VPN user holds which of these privileges (VPN user profile). An example for such privileges is the right to receive off-net to on-net calls or to establish on-net to off-net calls. When a call is set up for which a special privilege is required, the profile is checked to decide on the admission of the call.

- *Follow Me*: This function enables an on-net VPN user to install a temporary diversion to an arbitrary on-net or off-net location. For this purpose, the user connects to a special VPN service function to exchange the necessary information. Such a feature is extremely useful in a future world of complex multimedia services where a single user may have several services through which he or she communicates. It makes much sense just to tell the network about a diversion and not all the individual services.

- *Remote VPN Access*: Remote VPN access is used to make use of the VPN facilities from locations not identified as part of the VPN at the time of service provisioning. This can also be called a support function for 'roaming' users. In principle, the remote VPN access can be installed for a number of calls or only for one call. The Remote VPN access, installed for only one call, is sometimes also called 'through-dialing' through the VPN. As for a usual VPN call, the calling user identifies him-/herself as VPN member by his/her private number. However, the service logic will check if the terminal where the call is originated is an on-net location for the calling VPN user. If this is not the case, further information is needed to allow the calling user to place the call (e.g. a PIN).

## 2.4.1.2 Example 1: Admitted On-Net VPN Call

A call is denoted as an on-net VPN call, if a user who is a member of a VPN calls another user who is a member of the same VPN. During the call setup, the calling user provides his or her private number as Calling Party Number to identify him/herself as VPN member and the private number of the called user as Called Party Number. These private numbers trigger the invocation of the B-VPN service logic.

To ensure the integrity of the VPN, the VPN service logic has to check the membership of the users in the VPN (authentication) and the rights to place the required calls (authorization). After authentication and authorization, the service logic has to translate the private party number into a real network address to which the VPN call is to be routed. The VPN call is admitted if authentication, authorization and number translation was successfully performed by the service logic.

Figure 2.4.2 shows a typical network configuration for a VPN call which is the reference configuration for the exemplary scenario shown by the interaction diagram of Figure 2.4.3. In this scenario, the called user is not directly connected to the local exchange (the B-SSP) of the calling user. Therefore, NNI signaling is used to reach the called user, who is not explicitly depicted in Figure 2.4.3.

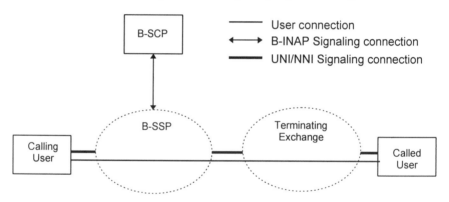

**Figure 2.4.2**
Network configuration for a VPN call

According to this scenario, the on-net VPN call proceeds as follows:

The user starts a call setup using private numbers, according to the private numbering plan of the same VPN subscription, as Calling Party Number (CgPN) and Called Party Number (CdPN) within the call SETUP message. These IN-numbers trigger DP 3 (see Section 2.2.4) as TDP-R within the CCF, causing the suspension of the call processing and a request towards the SSF. By this request, the SSF builds up an SSM instance as shown in Figure 2.4.4. This SSM instance is provided to the SCF (1).

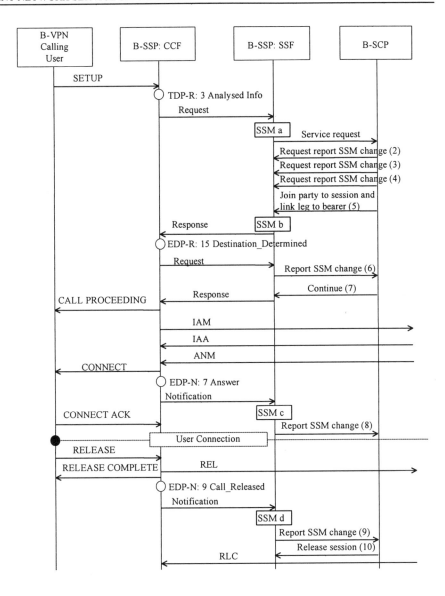

**Figure 2.4.3**
Establishment and termination of an admitted B-VPN call

*(1) Service request*

| | |
|---|---|
| Session ID | S1 |
| Service key | VPN service key |
| Calling party number | <VPN><VPN id><PN1> |
| Called party number | <VPN><VPN id><PN2> |
| SSM state: | *according to Figure 2.4.4* |

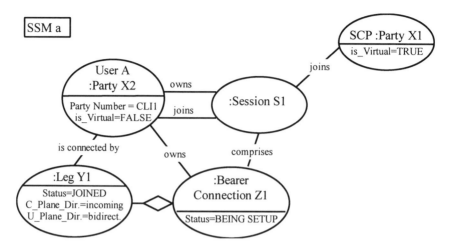

**Figure 2.4.4**
Object instances in the SSM at invocation of the IN functionality

It is assumed that IN-numbers for the B-VPN service are characterized by a specific prefix dialed by the user in order to access the service, indicated with <VPN>. The digits after <VPN> contain the VPN subscriber number (indicated as <VPN id>) and the private number of the VPN user (indicated as <PNn>). The physical network addresses of the involved parties are indicated as CLIn.

The service logic for the B-VPN service checks whether the CgPN and the CdPN appear in the subscription table of the indicated VPN as on-net locations of the VPN. Moreover, the service logic checks whether the network-supplied CLI1 corresponds to the calling user with the help of the subscription table for authentication and authorization purposes. After a successful authentication and authorization, the service logic translates the CdPN into the respective destination address, eventually making use of Time Dependent Routing (TDR) and Origin Dependent Routing (ODR).

Before providing the result of the translation to the SSP, the service logic requests further information about the evolution of the call (2)-(4).

*(2) Request report SSM change*
| | |
|---|---|
| Session ID | S1 |
| Object type | leg |
| *sSM Object ID list:* | |
|     SSM Object ID 1 | Y2 |
| *Object state list:* | |
|     *Object state 1:* | |
|         attribute Name | leg Status |
|         attribute Value | Destined |
|         monitor Mode | interrupted |
|         monitoring Info Requested | destinedLegInfo |
|     *Object state 2:* | |

|  |  |
|---|---|
| attribute Name | leg Status |
| attribute Value | Refused |
| monitor Mode | notifyAndContinue |

*(3) Request report SSM change*
| Session ID | S1 |
|---|---|
| Object type | leg |

*sSM Object ID list:*
| SSM Object ID 1 | Y1 |
|---|---|

*Object state list:*
*Object state 1:*
| attribute Name | leg Status |
|---|---|
| attribute Value | Abandoned |
| monitor Mode | notifyAndContinue |

*(4) Request report SSM change*
| Session ID | S1 |
|---|---|
| Object type | connection |

*sSM Object ID list:*
| SSM Object ID 1 | Z1 |
|---|---|

*Object state list:*
*Object state 1:*
| attribute Name | connection Status |
|---|---|
| attribute Value | Setup |
| monitor Mode | notifyAndContinue |

*Object state 2:*
| attribute Name | connection Status |
|---|---|
| attribute Value | Being released |
| monitor Mode | notifyAndContinue |

In order to enable a combination of the VPN-service with other IN services, it must be checked if the VPN-call is still admitted after the final routing information is available. For this, the leg status 'Destined' is monitored in the 'interrupted' mode to stop call processing until the service logic has performed this final check.

After these requests, the service logic provides the necessary information, in particular the destination address of the called user, to complete the call setup (5). This information extends the SSM as shown in Figure 2.4.5.

*(5) Join party to session and link leg to bearer*
| Session ID | S1 |
|---|---|
| Connection ID | Z1 |

*Bearer endpoint:*
| Party ID | X3 |
|---|---|
| Leg ID | Y2 |
| U Plane direction | Bi-directional |
| C Plane direction | Outgoing |
| Party number | CdPN |

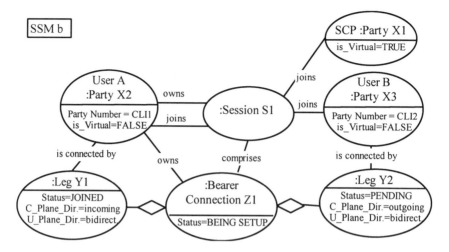

**Figure 2.4.5**
Object instances in the SSM after response from the service logic

If the final routing information is available, call processing is suspended again and the SCF is informed about this information (6).

*(6) Report SSM change*
    Session ID                       S1
    Object type                     leg
    sSM Object ID                 Y2
    *Object state:*
        attribute Name             leg Status
        attribute Value           Destined
        monitoring Info Reported   CLI2

Assuming that the VPN-call is still admitted, the SCF instructs the SSP to continue the call processing (7).

*(7) Continue*
    Session ID                       S1
    Object type                     leg
    sSM Object ID                 Y2

This message causes no change of the SSM instance. However, if the establishment of the connection was successful, this is detected at DP 7 and is reported to the SSF, resulting in the SSM instance as shown in Figure 2.4.6, and forwarded to the SCP (8).

*(8) Report SSM change*
    Session ID                       S1
    Object type                     connection

| sSM Object ID | Z1 |
|---|---|
| *Object state:* | |
| attribute Name | connection Status |
| attribute Value | Setup |

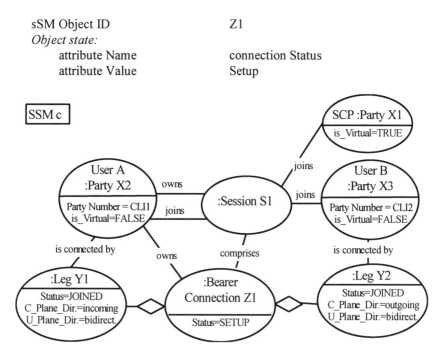

**Figure 2.4.6**
Object instances in the SSM after successful establishment of the connection

If one of the connected users finishes the call by sending a RELEASE message, the status of the Leg and Connection objects are modified accordingly, resulting in an SSM instance as shown in Figure 2.4.7, which is again reported to the SCP (9).

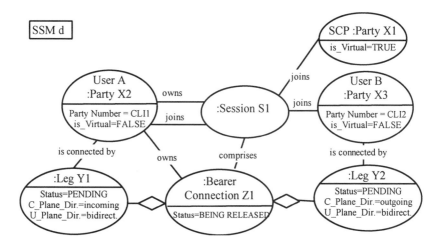

**Figure 2.4.7**
Object instances in the SSM after release of the connection

*(9) Report SSM change*
| | |
|---|---|
| Session ID | S1 |
| Object type | connection |
| sSM Object ID | Z1 |

*Object state:*
| | |
|---|---|
| attribute Name | connection Status |
| attribute Value | Being released |

Since the session is not used anymore by the SCF, the SCF releases the session (10). By receiving this message, the SSF disposes of the remaining objects belonging to that session.

*(10) Release session*
| | |
|---|---|
| Session ID | S1 |

## 2.4.1.3 Example 2: Usage of the Follow Me Function

The Follow Me function allows a VPN user to modify the entries of the VPN subscription tables. Of course, the user must be authorized for (eventually partial) modifications. One example is the modification of the associated network address to which a call to the VPN users should be routed. This example is presented in more detail in the following, using the USI mechanism (see Section 2.3.2) for the communication between user and service logic.

The interaction diagram of Figure 2.4.8 shows the messages exchanged between the involved network elements and/or functional blocks.

The user connects to the Follow Me service by establishing a bearer unrelated connection with the help of a COBI-SETUP message. The trigger mechanism with detection points is used to complete the connection via the SSF towards the SCF. Within the SSF, the SSM as shown in Figure 2.4.9 is created and, together with the USI information, provided to the SCF (1).

*(1) Service request*
| | |
|---|---|
| Session ID | S1 |
| Service key | VPN service key |
| CallingPartyNumber | <VPN><VPN id><PN1> |

USI information:
| | |
|---|---|
| Service Identifier | B-VPN Service |
| Operation Identifier | FollowMeDestinationMod.req.ind |
| Destination Address | CLI3 |
| PIN | xyz |
| SSM state: | *according to Figure 2.4.9* |

In contrast to example 1 (see Subsection 2.4.1.2), only the status of connections and not of legs are monitored. Since the connection has already the status 'Setup', only the status change to 'Being released' of the connection is requested to be reported (2).

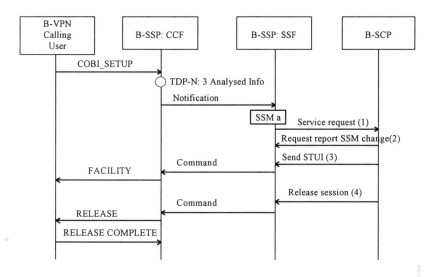

**Figure 2.4.8**
Invocation of the Follow Me function

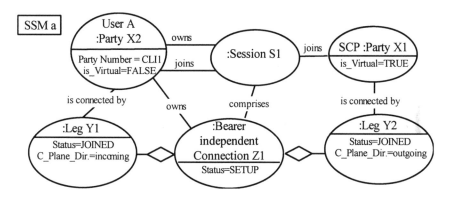

**Figure 2.4.9**
Object instances in the SSM for an invocation of the Follow Me function

*(2) Request report SSM change*

| | |
|---|---|
| Session ID: | S1 |
| Object type | connection |
| sSM Object ID list: | |
|     SSM Object ID | Z1 |
| Object state list: | |
|     Object state 1: | |
|         attribute Name | connectionStatus |
|         attribute Value | Being released |
|         monitor Mode | notifyAndContinue |

If the service logic accepts the request, it checks afterwards whether the user is authorized for using the Follow Me function by comparing his/her CLI and the PIN with the information contained in the VPN subscription table. After a plausibility check, the VPN database is updated. If the checks were successful, the SCP sends the required USI information, which in this case is a positive response, to the SSF (3) which forwards it via the CCF towards the user.

*(3) Send STUI*
    Session ID:                  S1
    Leg ID                    Y2
    USI information:
        Service Identifier         B-VPN Service
        Operation Identifier     FollowMeDestinationMod.resp.conf
        Return Code           NoError - RequestAccepted

Since the Follow Me function requires only a simple request/response dialogue, the SCF releases the session directly after sending the response (4).

*(4) Release session*
    Session ID                  S1

## 2.4.2   Interactive Multimedia Retrieval

Interactive Multimedia Retrieval (IMR) is a complex service which fully exploits the flexibility in connection allocation and control provided by the Intelligent Broadband Network. Especially the handling of more than one connection within one session and the use of SCP-initiated calls can be demonstrated by this service.

After an introduction into the IMR service and the network configuration supporting this service, a complete scenario of an IMR session is presented.

### *2.4.2.1 Introduction*

The IMR service provides the user with the means to select a multimedia application and to retrieve audio-video information stored in information centers. The information will be sent to the user on his demand only and can be retrieved on an individual basis. The user controls the time at which an information sequence is to start. Using IN functionality, a flexible broking facility among available service providers can be provided for the service user. Three types of contents are provided within the general IMR framework, that is, Video On Demand (VOD), News On Video (NOV) and Karaoke/Music On Demand (KOD/MOD).

Service provision in this example service is fully based on ATM, end-to-end. The user employs a Set-Top Box (STB) in order to perform the necessary service selection and view the desired contents. A Video Server (VS), from which the download of digital contents is performed, is located in the service provider premises. The involved network elements are a B-SSP, possibly several ATM switches, a B-SCP and a B-IP. Figure 2.4.10 shows the IMR network reference configuration.

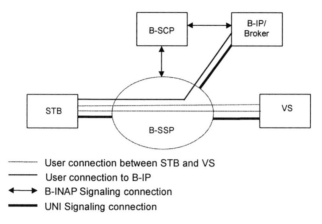

--------- User connection between STB and VS
————— User connection to B-IP
◄——► B-INAP Signaling connection
————— UNI Signaling connection

**Figure 2.4.10**
IMR reference configuration

The service makes use of several advanced signaling and IN capabilities as described in Chapter 2.3. From the signaling viewpoint, the following ones are noteworthy:
- UNI and NNI bandwidth negotiation [3, 4];
- UNI and NNI bandwidth modification [5, 6].
  From the IN viewpoint, IMR makes use of the following features:
- Number translation;
- Service provider selection;
- Authentication/authorization;
- SCP-initiated call;
- Multiple connections within a single service session.

## 2.4.2.2 IMR as an Intelligent Broadband Network service

The service is composed of two different phases.

During Phase 1, the user selects the IMR service by dialing the appropriate IN number and is connected by IN to the B-IP, where authentication/authorization and service provider selection take place. After the completion of Phase 1, the Intelligent Broadband Network disconnects the user from the B-IP and reconnects him with the selected server.

At this point, the user selects the desired content and the playback of the digital stream begins. Interaction can take place in the form of commands for stream control (stop, fast forward, ...).

When playback ends, the IN reconnects the user to the B-IP, where he can choose another service provider or disconnect from the service as well.

The Figures 2.4.11 to 2.4.16 report the information flows exchanged between the different elements taking part in providing the IMR service during the two phases.

## Phase 1, Step 1
The SETUP message sent by the STB contains the IMR number which triggers IN at TPD-R 3 (Analyzed Info).

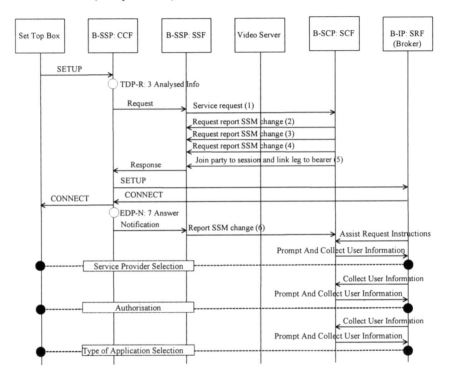

**Figure 2.4.11**
Phase 1, Step 1: Connection of the STB to the B-IP

The B-SSP reacts by sending a 'Service request' message towards the B-SCP asking for the establishment of an IMR session. On acceptance of the request, the B-SCP instructs the B-SSP in order to monitor state changes of SSM objects (by means of 'Request report SSM change' messages) and gives the B-SSP routing information in order to reach the B-IP (Join party to session and link leg to bearer). The B-SSP is now able to route the call to the B-IP (SETUP message). The reception of the CONNECT message coming from the B-IP is reported to the B-SCP in the form of a 'Report SSM change' message (EDP-N 7: Answer Notification). The User Plane dialogue between the user and the B-IP is reported to the B-SCP which in turn instructs the B-IP on how to proceed (exchange of 'Assist Request Instructions'/'Prompt And Collect User Information'/'User Information' messages).

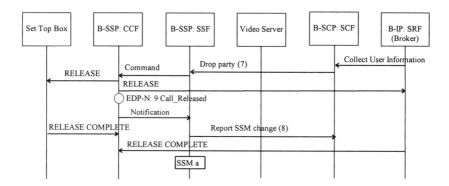

**Figure 2.4.12**
Phase 1, Step 2: Release of the call towards the B-IP

*Phase 1, Step 2*

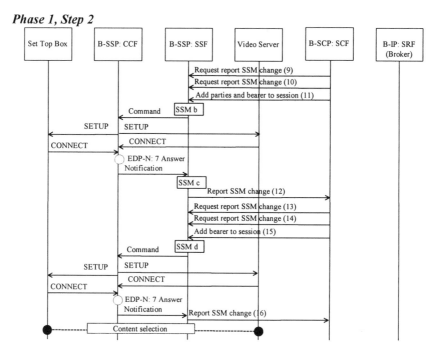

**Figure 2.4.13**
Phase 2, Step 3: Connection of the STB to the selected Video Server

The B-SCP detects the end of the selection dialog by means of the last 'User Information' message coming from the B-IP and instructs the B-SSP in order to release the connection between STB and B-IP ('DropParty' message). This action is translated into RELEASE signaling messages and bearer release is reported to the B-SCP accordingly ('Report SSM change' message, EDP-9: Call_Released).

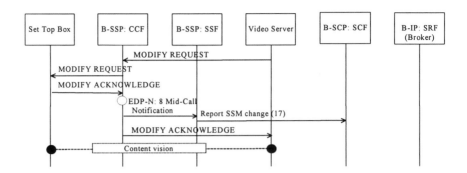

**Figure 2.4.14**
Phase 2, Step 4: Bandwidth modification according to the selected contents

*Phase 2, Step 3*
The B-SCP now asks the B-SSP to monitor state transitions of objects involved in subsequent SCP-initiated calls. Connections between the STB and the selected video server are created by the B-SSP on reception of appropriate messages from the B-SCP (Add parties and bearer to session, Add bearer to session). State changes are reported accordingly.

*Phase 2, Step 4*

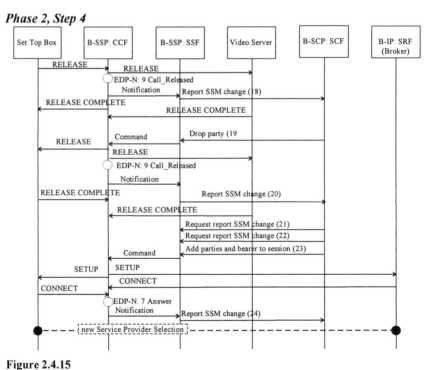

**Figure 2.4.15**
Phase 2, Step 5: Release of the calls to the Video Server and reconnection to the B-IP

The user is now connected to the video server and is free to select a video content. After the selection, the video server modifies the bandwidth of the connection (which was initially set to 0 by the B-SCP when creating the connection itself) in order to reflect the actual needed value, which obviously depends on the user's selection. A signaling MODIFY message is used. The acceptance of bandwidth modification by the STB (MODIFY ACKNOWLEDGE) is reported to the B-SCP in order to keep track of the actual bandwidth of the connection.

### Phase 2, Step 5
When the user wants to disconnect from the video server, a RELEASE message is sent. This affects one of the two bearers and is reported to the B-SCP (EDP-N 9: Call_Released Notification). The remaining connection is released after a 'Drop party' message sent by the B-SCP. SSM state changes are reported to the B-SCP, which is now able to reconnect the user to the B-IP by sending an 'Add parties and bearer to session' message.

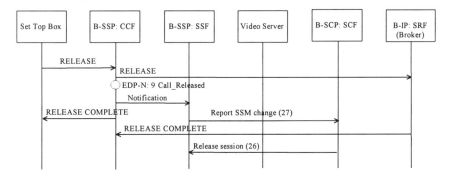

**Figure 2.4.16**
Phase 2, Step 6: Release of the call towards the B-IP - end of session

### Phase 2, Step 6
The user is free to select a (new) video server or to release the call, as shown in the Figure 2.4.16. The release is notified to the B-SCP which closes the IMR session by sending a 'Release session' message.

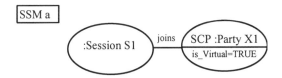

**Figure 2.4.17**
SSM instance after clearing of the STB/B-IP connection

As an example of the employed information flows and related SSM evolution, the case of the two SCP-initiated calls connecting the STB and the VS is presented (Phase 2, step 3). The service session is considered as already established according to the previously described IMR phases. Messages are numbered according to Figure 2.4.13.

After the release of the connection between the B-IP and the STB the SSM state is as it appears in Figure 2.4.17.

The B-SCP now connects the STB and the desired service provider. This is accomplished by means of two SCP-initiated calls.

The first call is used in order to establish the control connection between STB and VS (11), which is bi-directional and low bit rate. The report of SSM state changes is armed in advance (9)-(10).

*(9) Request report SSM change*
| | |
|---|---|
| Session ID: | S1 |
| Object type | connection |
| *Object ID list:* | |
|     Object ID | Z2 |
| *Object state list:* | |
|     *Object state 1:* | |
|         Attribute | connection Status |
|         Attribute value | Setup |
|         Monitor mode | notifyAndContinue |

*(10) Request report SSM change*
| | |
|---|---|
| Session ID: | S1 |
| Object type | leg |
| *Object ID list:* | |
|     Object ID | Y3,Y4 |
| *Object state list:* | |
|     *Object state 1:* | |
|         Attribute | leg Status |
|         Attribute value | Refused |
|         Monitor mode | notifyAndContinue |
|     *Object state 2:* | |
|         Attribute | leg Status |
|         Attribute value | Pending |
|         Monitor mode | notifyAndContinue |
|         Monitoring Info Req. | being Released Info |

*(11) Add parties and bearer to session*
| | |
|---|---|
| Session ID | S1 |
| *Bearer end point list:* | |
|     *Bearer endpoint 1:* | |
|         Party ID | X4 (Set Top Box party) |
|         Leg ID | Y3 |
|         U plane direction | Bi-Directional |
|         C plane direction | Outgoing |
|         Party number | E.164 number of X4 |

|  |  |
|---|---|
| High layer information | service-specific content |
| Low layer information | service-specific content |
| *Bearer endpoint 2:* |  |
| Party ID | X5 (Video Server party) |
| Leg ID | Y4 |
| U plane direction | Bi-Directional |
| C plane direction | Outgoing |
| Party number | E.164 number of X5 |
| High layer information | service-specific content |
| Low layer information | service-specific content |
| Connection ID | Z2 |
| *Bearer characteristics:* |  |
| Forward Peak cell rate | 167 cell/sec |
| Backward Peak cell rate | 167 cell/sec |
| Bearer class | BCOB-X |
| AAL type | AAL5 |
| Connection owner ID | X5 |

The SSM instance goes into a state which is shown in Figure 2.4.18.

After the reception of the two CONNECT messages coming from STB and VS, the B-SSP reports the successful setup to the SCP (12).

*(12) Report SSM change*

|  |  |
|---|---|
| Session ID | S1 |
| Object type | connection |
| Object ID | Z2 |
| *Object state:* |  |
| Attribute | connection Status |
| Attribute value | Setup |

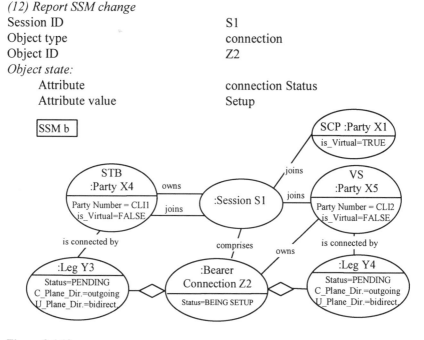

**Figure 2.4.18**

Object instances in the SSM after the instruction to connect the STB with the Video Server

The SSM instance is now structured as Figure 2.4.19 shows.

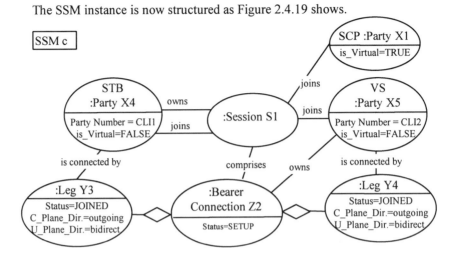

**Figure 2.4.19**
SSM instance after the successful completion of the first SCP-initiated call

The B-SCP now commits the establishment of a second bearer between STB and VS (15), used for the unidirectional (VS to STB direction) transmission of a digital stream. This connection is established at zero bandwidth and will be subsequently modified by the VS (by means of the bandwidth modification procedure, see Chapter 2.3) according to the bandwidth requirements of the content selected by the user.

State change reports are again requested (13)-(14).

*(13) Request report SSM change*
| | |
|---|---|
| Session ID: | S1 |
| Object type | connection |
| *Object ID list:* | |
|    Object ID | Z3 |
| *Object state list:* | |
|    *Object state 1:* | |
|       Attribute | connection Status |
|       Attribute value | Setup |
|       Monitor mode | notifyAndContinue |
|       Monitoring Info Requested | setup Bearer Info |

*(14) Request report SSM change*
| | |
|---|---|
| Session ID: | S1 |
| Object type | leg |
| *Object ID list:* | |
|    Object ID | Y5,Y6 |
| *Object state list:* | |
|    *Object state 1:* | |
|       Attribute | leg Status |

| | |
|---|---|
| Attribute value | Refused |
| Monitor mode | notifyAndContinue |
| *Object state 2:* | |
| Attribute | leg Status |
| Attribute value | Pending |
| Monitor mode | notifyAndContinue |
| Monitoring Info Requested | being Released Info |

*(15) Add bearer to session*

| | |
|---|---|
| Session ID | S1 |
| *Bearer end point list:* | |
| *Bearer endpoint 1:* | |
| Party ID | X4 (set Top Box party) |
| Leg ID | Y5 |
| U plane direction | Bi-Directional |
| C plane direction | Outgoing |
| High layer information | service-specific content |
| Low layer information | service-specific content |
| *Bearer endpoint 2:* | |
| Party ID | X5 (Video Server party) |
| Leg ID | Y6 |
| U plane direction | Bi-Directional |
| C plane direction | Outgoing |
| High layer information | service-specific content |
| Low layer information | service-specific content |
| Connection ID | Z3 |
| *Bearer characteristics:* | |
| Forward Peak cell rate | 0 cell/sec |
| Backward Peak cell rate | 0 cell/sec |
| Bearer class | BCOB-X |
| AAL type | AAL5 |
| Connection owner ID | X5 |

The SSM instance now contains new objects in a transient state, as Figure 2.4.20 shows.

As it was the case for the previous SCP-initiated call, after the reception of the two CONNECT messages coming from STB and VS, the B-SSP reports the successful setup to the SCP (16).

*(16) Report SSM change*

| | |
|---|---|
| Session ID | S1 |
| Object type | connection |
| Object ID | Z3 |
| *Object state:* | |
| Attribute | connection Status |
| Attribute value | Setup |

All the SSM objects have now reached a steady state (as in Figure 2.4.20) with the status of all legs being 'Joined' and the status of all bearer connections being 'Setup'.

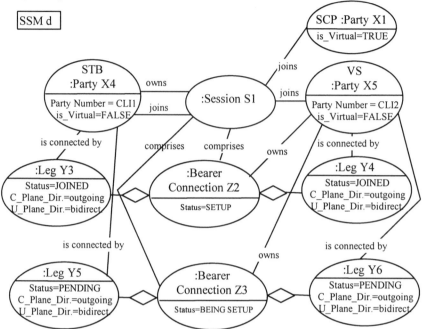

**Figure 2.4.20**
Object instances in the SSM after the instruction to establish the second bearer between the STB and the Video Server

## 2.4.3   Broadband Video Conference

The Broadband Video Conference (B-VC) service is the most complex service of the three services presented in this chapter. However, no new IN/B-ISDN features compared with Sections 2.4.1 and 2.4.2 are used by this service. Therefore, detailed scenarios, which would expand the book without providing much new information, are omitted. Additional information about the realization of the B-VC service within the INSIGNIA project can be found in [7].

Instead, the advantages of the Intelligent Broadband Network approach for the B-VC service are discussed, emphasizing the distribution of intelligence between application and network. Furthermore, several alternatives for the realization of the B-VC service using the new IN/B-ISDN features are sketched.

### 2.4.3.1 Introduction

The B-VC service is an example of a complex multimedia communication service. It supports more than two parties and several media types (e.g. video, audio, control, whiteboard data). Therefore, it is an ideal example to illustrate the usage of the advanced signaling capabilities arising from the IN/B-ISDN integration. Figure 2.4.21 shows a typical connection configuration of the B-VC service.

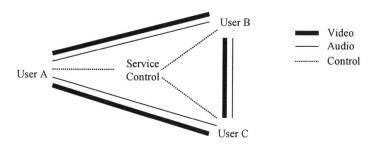

**Figure 2.4.21**
Typical B-VC connection configuration

Figure 2.4.21 illustrates that there is a need to coordinate at least nine connections of three different types in order to run a video conference among three parties. There are fully meshed audio and video connections as well as a control connection from each party towards a central service control instance, which is identical to the SCP in a true IN realization of the service.

When the B-VC service is applied, many advantages of the Intelligent Broadband Network approach become even more visible than for other examples. These advantages are:

- The user is able to use the features of the video conference in a very abstract way. This means that the user just fills in a form denoting the conferees and their roles and rights to describe a conference. It is the task of the Intelligent Network to translate such an abstract conference description into an actual invocation sequence of connection establishments.

- The level of abstraction introduced by the Intelligent Network helps in providing a smooth service migration towards improved networks. For example, the same service as shown in Figure 2.4.21 can be offered to the user by using point-to-multipoint connections as shown in Figure 2.4.22 [8].

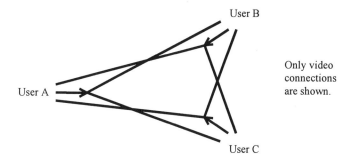

**Figure 2.4.22**
Alternative B-VC connection configuration

- This realization alternative optimizes the usage of network resources for larger conferences. A transition from a simple (B-ISDN CS-1) network to a more advanced network may take place without bothering the users at all.
- The fact that the complex connection configuration for a single instance of the B-VC service is kept together in a single IN session is of great help for administrative purposes. For example, it is easy to define billing schemes for video conferences based on IN. For instance, the initiator of a conference may have to pay for the whole conference or the cost may be evenly distributed among the conferences. In any way, the IN sessions are the central glue keeping together the various pieces (connections) of the service.

## 2.4.3.2 Video Conference Functions

The B-VC service is not a single chain of control, but rather a collection of individual functions. The most important functions are as follows.
- *Conference creation*: A user of the service creates a new description of a conference (conference record). Many details on the conference can be entered, but there is no effect on actual communication connections yet. The conference is still inactive;
- *Conference establishment*: A user triggers the initiation of a communication session according to a previously stored conference record. The conferees are invited by the Intelligent Network to participate, and the required connection configuration is built up between those conferees who accept the invitation. The conference goes active;
- *Conference modification*: A user modifies the contents of a conference record, e.g. to modify the permissions of a conferee. This operation can be applied to active and inactive conferences. For active conferences, a number of specific modification functions can be distinguished which cause immediate changes to the connection configuration. Examples are to add a new conferee or to disconnect a conferee;
- *Join conference*: A user requests to be included into an active conference. For this purpose, he or she gives a command to the Intelligent Network identifying a specific conference. The IN takes care of asking the other conferees for permission and updating the communication configuration;
- *Leave conference*: A user disconnects him or herself from an active conference;
- *Close down conference*: A user, who has the appropriate permission, terminates an active conference. The conference goes inactive;
- *Delete conference*: A user deletes an inactive conference record.

The B-VC service is defined by its *conference state*, *conferees* and all established *connection sets* between each pair of conferees: user data connections for audio, video and data transfer and connections between each CPE and a central conference control function. The conference description (topological definition, attributes and state of a certain conference, of the conferees and the connection sets) is stored in the Conference Information Record (CIR) within the

Conference Information Data Base (CIDB), located within and used by the central conference control function.

Two level of services are offered: a *basic conferencing service* and a more sophisticated 'high level' conferencing, which includes a number of different conference policies called *Conference Models*. The basic service exclusively uses distributed functionality, thus avoiding the introduction of a single point of failure.

## 2.4.3.3 B-VC as an Intelligent Network Service

In order to realize the B-VC service on an Intelligent Network infrastructure, it has to be decided how the functionality mentioned above is distributed over the Physical Entities (PEs) of this network architecture. Such a reference architecture for B-VC service deployment in terms of Physical Entities (PEs) is given in Figure 2.4.23. It includes Customer Premises Equipment's (CPEs)[9], B-ISDN switching systems and IN components.

The CPE is a computer device serving as an endpoint for the audio and video connections. It provides network access and a graphical user interface to the video conference software. Audio and video multiplexing is done at the CPE. Each user having access to a CPE can take part in a conference as a conferee.

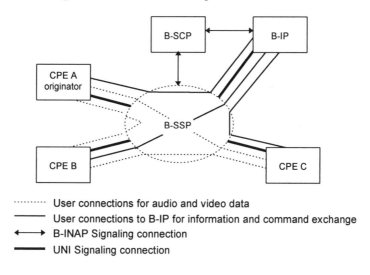

                    ········· User connections for audio and video data
                    ──────── User connections to B-IP for information and command exchange
                    ◄───► B-INAP Signaling connection
                    ──────── UNI Signaling connection

**Figure 2.4.23**
Broadband Video Conference reference architecture

The central conference control function, which was mentioned above, can be realized in a way that mirrors the two-level structure of the service. Basic conference control functions are realized by the B-SCP. For more sophisticated conferencing, a B-IP is employed in order to handle user interactions. It plays the role of a special Conference Server and implements the logic which is in charge of a High Level Conference Management. This includes:

- Performing advanced communication with users;
- Acting as a relay for communication between B-SCP and users;
- Holding the Conference Information Database (or providing access to it);
- Management of static aspects of conferences;
- Providing basic services to the Conference Models;
- Running the Conference Models.

A connection between B-IP and B-SCP is needed in order to allow the transfer of user choices performed on the B-IP.

A specification of the Intelligent Network implementation of B-VC has to map all the functions from above to the B-INAP protocol and to define message sequence charts for all the functions.

There are several implementation alternatives for a complex service like B-VC on a B-IN infrastructure. Some alternatives are discussed briefly below, restricting attention to the conference establishment function.

One possibility is to maintain for every single conferee a low-bit rate bearer connection to the B-IP as a means to exchange status and control information. In this case, the first steps of a conference establishment look like the first phase of the IMR service. The initiating conferee establishes a connection by an IN number which is translated into the address of the B-IP. Afterwards, a dialogue between B-IP and B-SCP is carried out to transfer conference record information, and the users are invited by SCP-initiated call setups. Technical details are omitted here.

A more interesting variant tries to avoid the bearer connection between user and B-IP and uses User-Service Interaction for control of the conference. In this case, the initiating user sends a COBI-SETUP which leads to a bearer-unrelated connection to the SCP analogous to the Follow Me function for the VPN service (see Figure 2.4.8). However, instead of releasing the session, the invitation and connection setup phases take place after the User-Service Interaction.

For the invitation and connection setup phases, again several implementation alternatives are possible. If the feature of an SCP-initiated setup of a bearer-unrelated connection is available, the invitation can be delivered using User-Service Interaction. Otherwise, it is also possible to combine connection setup phase and invitation phase and to signal the invitation to the software on the conferees' terminals by SCP-initiated bearer connections between the conferees.

To summarize, Broadband IN provides a rich set of features which turns out to be sufficient for classical telephone-style IN services as well as for very advanced and complex multimedia services.

# References

[1] von der Straten G, Totzke J and Zygan-Maus R, *Realization of B-IN Services in a Multi-national ATM Network*, Proceedings of 4th International Conference on Intelligence in Networks (ICIN'96), Bordeaux, France, November 1996, pp. 41-46

[2] Ambrosch W D, Maher A and Sasscer B, *The Intelligent Network*, Springer, 1989

[3] ITU-T Recommendation Q.2962, *DSS2 - Connection characteristics negotiation during call/connection establishment*, 1997

[4] ITU-T Recommendation Q.2725.1, *B-ISDN User Part - Support of negotiation during call setup*, 1996

[5] ITU-T Recommendation Q.2963.1, *DSS2 - Peak Cell Rate modification by the connection owner*, 1996

[6] ITU-T Recommendation Q.2725.2, *B-ISDN User Part - Modification*, 1996

[7] Brandt H, Tittel C and Todorova P, *Broadband Video Conference in an Integrated IN/B-ISDN Architecture*, Proceedings of European Conference on Networks and Optical Communication (NOC'97), Antwerp, Belgium, June 1997

[8] Listani M, Salsano S, *Point-to-multipoint Call Modeling for the Integration of IN and B-ISDN*, Proceedings of 2IN'97, Paris, France, September 1997

[9] Brandt H, Tittel C, Todorova P, Tchouto J J and Welk M, *Broadband Video Conference Customer Premises Equipment*, Proceedings of European Workshop on Interactive Distributed Multimedia Systems and Telecommunication Services (IDMS'97), Darmstadt, Germany, September 1997

# Chapter 2.5

# THE BROADBAND SERVICE SWITCHING POINT - B-SSP

The functionality of Intelligent Broadband Networks can be designed starting from the principles underlined in the previous chapters. A particular advantage of this approach is that it can be integrated in existing systems without great impact on already available structures.

The focus of this chapter is on the enhancements imposed on ATM switching systems in order to perform trials with Intelligent Broadband Networks. Experiments allow to verify the adequacy of the adopted design solutions to the problem of providing advanced IN services in a Broadband context.

General design issues are discussed in Section 2.5.1. In Section 2.5.2, specific platforms are outlined which were taken as a basis for the implementation of B-SSPs in the INSIGNIA project. The task of embedding Service Switching functionality into signaling software is discussed in Section 2.5.3 with the help of three study cases related to specific B-SSP implementations.

## 2.5.1 Common Object-Oriented Design Models

The models achieved during the analysis of the system (see Chapter 2.2) provide a starting point for the development of the design models. However, some aspects common for all B-SSPs are not specified in Chapter 2.2, in particular a model for the CCF/CUSF, a model for the data to be administered and the relationships between the classes of the Functional Units.

This additional information is provided in the following, building the common basis for the special design of the different B-SSPs. The common design consists of models for each of the components which must be newly developed for B-SSPs supporting the new IN/B-ISDN concepts, that is the functional entities of a Service Switching Function (SSF) and a Call Control/Call Unrelated Service Function (CCF/CUSF) as well as a local component of the B-SSP which provides the special administrative features needed by IN, and the relationship between these components.

As an extension of the models described in Chapter 2.2, the common design models are also developed according to the object-oriented methodology. As proposed by OMT (see [1] and Appendix B), the static structure of the identified objects is described by class diagrams. The dynamic behavior of the objects is expressed by interaction diagrams and state diagrams.

*Intelligent Broadband Networks*. Edited by I. S. Venieris, H. Hussmann
© 1998 John Wiley & Sons Ltd.

### *Call Control Function/ Call Unrelated Service Function*

The main task of these functional units is to handle the UNI and NNI signaling protocol and to establish the requested connections. Since the main difference between these functional units is the kind of connections which they control (respectively bearer connections and bearer independent connections), one design model for them can be provided.

In addition to the 'normal' call processing for switched connections, the following additional features are realized for IN/B-ISDN:

- Establishment of SCP-initiated calls;
- SCP-initiated release of calls;
- Realization of the BCSM to provide an interface for the interaction with the SSF;
- Bandwidth negotiation and modification;
- User-Service Interaction.

The CCF model for B-SSPs is derived from call processing models of ordinary B-ISDN switches (see, for example, [2]). In the following, an extension of such models is described emphasizing the realization of the new features. The necessary adaptations for the interaction with the SSF can be derived from the SSF model.

An *SCP-initiated call* is a means to setup a switched connection between two users, initiated by the B-SCP. With respect to signaling, the two users take the role of 'called users', whereas the roles of the corresponding 'calling users' are played by the network. This feature is similar to the B-ISDN feature 'third party call', where a third user establishes a switched connection between two other users.

To provide the feature of an SCP-initiated call, the object model of the CCF can be extended as shown in Figure 2.5.1.

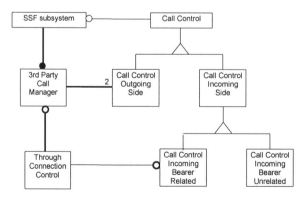

**Figure 2.5.1**
Object model for the SCP-initiated call and call unrelated control

An SCP-initiated call is under the control of a *3rd Party Call Manager*. Whenever the B-SCP wants to establish a connection between two users, this 3rd Party Call Manager is involved. At first, the 3rd Party Call Manager initiates the call setup like two calling users, simulating the objects of the 'Incoming Side' so that the 'Outgoing Side' objects will not be affected. Therefore, it does not matter if the corresponding real users, playing the role of called parties, are connected via UNI or NNI. This behavior is modeled by the association between the 3rd Party Call Manager and the class Call Control Outgoing Side. As soon as the 3rd Party Call Manager gets reports of the call acceptance from both involved call control instances, it initiates the through-connection of the bearer. After the through-connection is reported as successful, the 3rd Party Call Manager forwards this information to both involved Call Control Outgoing Side objects, which resume call processing as usual.

The release of a call, initiated by the SCP, is performed analogously. Additionally, the 3rd Party Call Manager must be informed of a call release, sent by one of the called parties, which can happen at any time including the setup phase. These events result in the state transitions shown in Figure 2.5.2.

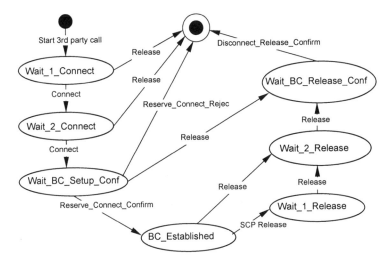

**Figure 2.5.2**
State diagram of the 3rd Party Call Manager

The Call Unrelated Control can be treated as a specialization of the class Call Control Incoming Side, because it is based on the communication of one user with the network. The main difference between the classes Bearer Related and Bearer Unrelated Call Control is the capability to establish a bearer connection which is expressed by an association between Bearer Related Call Control and the Through-Connection Control, but which is missing for the Bearer Unrelated Call Control.

### Service Switching Function
The SSF provides features to support the IN-services for Intelligent Broadband Networks. In particular, the following tasks must be realized by the SSF:

*1. Detection Point Handling:*
- Management of Event Detection Points;
- Check if a Detection Point is armed;
- Communication with the Session management.

*2. Session Management*
- Invocation of a third party call as SCP-initiated call in the CCF;
- Initiation of the release of a call (SCP-initiated call release);
- Communication with the corresponding Call Control objects (for third party calls, plain user-to-user calls and bearer independent calls) belonging to the session;
- Translation of BCSM-states into SSM-states and vice versa;
- Arming Event Detection Points;
- Communication with the SCF Access Manager.

*3. Communication with the B-SCP*
- Interpretation of the B-INAP-messages;
- Communication with the TCAP-Handler;
- Control of the information flow between the B-SSP and the B-SCP;
- Information of the B-SCP about the SSM state;
- Receipt of B-SCP commands and B-SCP responses.

The core of the SSF design model is the SSM as described in Chapter 2.2. The SSM classes can be used directly. The only task left to do is to bring these classes into relationship with other software components. The class diagram in Figure 2.5.3 shows the interface classes of the CCF/CUSF and the SCF with respect to the session, the central class of the SSF.

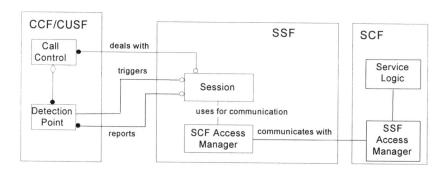

**Figure 2.5.3**
Object model for the SSF

To each *Call Control*, representing one basic half-call, a BCSM containing *Detection Points* is assigned. On reaching an armed Trigger Detection Point (TDP) during call processing, a *Session* object is created. Usually, one detection point can trigger only one IN Service, represented by a Session. If a combination of IN services is to be provided, it may be necessary that a detection point triggers several services/sessions. However, the model can be extended accordingly by changing the multiplicity of the 'triggers' association at the 'Session' end to 'many'. Each Session is in one-to-one correspondence with an *SCF Access Manager*. Communication between SCF and SSF is represented by the one-to-one association between *SSF Access Manager* and SCF Access Manager.

The associations 'triggers' and 'deals with' between CCF and SSF objects are managed by the functional block *Basic Call Manager*, implementing the DP processing, and by the Functional Unit *IN Switching Manager* (see Figure 2.2.2). The IN Switching Manager is responsible for the handling of the SSM objects Session, Party, Leg and Connection, in particular the creation, deletion and updating of these objects.

### Local IN-Management

The administrative features for a B-SSP are concerned with the management of IN-numbers, trigger detection points, routing and subscriber information and the administration of routing towards reachable B-SCPs. The object model shown in Figure 2.5.4 provides an abstract view of the local IN management of the B-SSP.

This model describes the administration capabilities which are specific to the IN extensions of the B-SSP. In other words, the given model depicts the 'managed objects' which are needed in addition to management for an ordinary B-ISDN switch due to the Intelligent Broadband Network. From this model, it is possible to derive suitable commands offered to the administrator of the B-SSP.

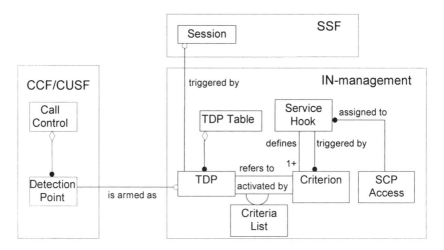

**Figure 2.5.4**
Object model of the IN management

In the list below, the meaning of each class is sketched briefly:

- The *TDP Table* contains the trigger detection points which must be statically administrated.

- The *TDP class* describes a single TDP that can trigger IN involvement. It is possible to assign criteria to a TDP. One of these criteria must be met in order to trigger an IN-service.

- The TDP *criteria* are placed in priority order by means of administrative procedures in a *Criteria_List*. This priority affects the association between the TDP class and the Criterion class.

- A *Criterion* is a generalization of the criteria to which a TDP is related. More specialized criteria could be the IN-prefix of a calling number and the IN-prefix of a called number.

- A *Service Hook* represents the interface between the service logic of an individual IN service processed by a B-SCP and the basic trigger mechanism of the call processing. The interface to the basic call processing consists of one or more criteria which constitute the trigger mechanism of the service and allow the basic call processing to interact with the service logic.

- The *SCP Access* defines the way in which the B-SCP is accessed. This class represents the capability to administer the port of the B-SSP by which the B-SCP is accessed.

## 2.5.2   Switching Platforms as Basis for Implementing B-SSPs

This section gives three detailed examples for ATM switching systems which were employed in the INSIGNIA project as platforms to build B-SSPs.

### 2.5.2.1 Italtel Platform

The Italtel B-SSP has been built around the concept of an external server which, being connected to an ATM cross-connect (see [3, 4]) by means of a standard ATM interface, upgrades it to a switching system. In INSIGNIA, the server has been provided with IN capabilities as well, so that the complete system behaves as a Broadband Service Switching Point. This implementation of the B-SSP is therefore divided into two subsystems, as Figure 2.5.5 shows.

Users and network devices are connected to ATM ports of the subsystem 'ATM switching platform', which can support electrical E3, optical SDH/STM1 and TAXI interfaces. The subsystem 'Signaling/IN server' is connected to a dedicated port of the ATM switching platform. Signaling and IN flows are collected from the various ports and sent to the Signaling/IN server by means of cross-connections which are pre-established via management operations (see [5]). The server distinguishes the ports from which signaling/IN information comes from by means of the ATM channel which carries it, elaborates the received flows and instructs the switching matrix in order to establish or release connections between the ports, according to the operations performed in the Control Plane. Figure 2.5.6 illustrates the relationships between the two subsystems.

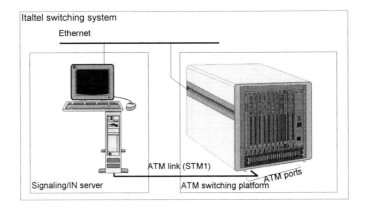

**Figure 2.5.5**
Hardware structure of the Italtel switching system

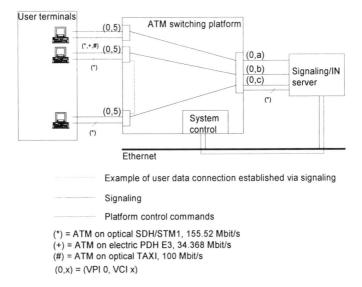

Example of user data connection established via signaling

Signaling

Platform control commands

(*) = ATM on optical SDH/STM1, 155.52 Mbit/s
(+) = ATM on electric PDH E3, 34.368 Mbit/s
(#) = ATM on optical TAXI, 100 Mbit/s
(0,x) = (VPI 0, VCI x)

**Figure 2.5.6**
Relationships between the two subsystems of the Italtel switching system

In this particular implementation, instructions to the ATM cross-connect are sent via its management channel, which is based on SNMP carried over Ethernet. A particular software module of the server ('Resource Manager') is responsible for the control of the ATM system. By means of adaptations to this block, the Signaling/IN server can be used to control several types of ATM cross-connects leaving the rest of the software unchanged. The server approach provides therefore a high flexibility which is particularly needed in case of experimental usage of ATM switches.

The ATM system used in the Milan site of the INSIGNIA trial is the UT-ASM developed within the framework of the RACE project R2118 (BRAVE). In the Turin site of the INSIGNIA trial, the same kind of server is used to control a FORE systems ASX200E cross-connect.

The Signaling/IN server of the first INSIGNIA experimental phase is based on a Sun SPARC20 workstation running Solaris 2.5. For the second phase of the INSIGNIA experimentation, evolution towards real-time OS and PowerPC processor is being performed.

The Italtel platform offers the following functionality:

- Optical SDH/STM1, electrical E1, optical TAXI physical interfaces;
- ITU-Q.2931/ATM Forum 3.1 (UNI) and ITU-T Q.2762-Q.2764 (NNI) signaling;
- ITU-T Q.2932/COBI (for user/service interaction);
- ITU-T Q.2962 (bandwidth negotiation at the UNI during call setup);
- ITU-T Q.2725.1 (bandwidth negotiation at the NNI during call setup);
- ITU-T Q.2963.1 (bandwidth modification at the UNI);
- ITU-T Q.2725.2 (bandwidth modification at the NNI);
- Resource Manager for the control of the ATM switching platform;
- SNMP configuration and alarms handling.

The following figures describe the functional organization of the Signaling/IN server software. Figure 2.5.7 shows the overall architecture while Figure 2.5.8 focuses on the CCF internal architecture.

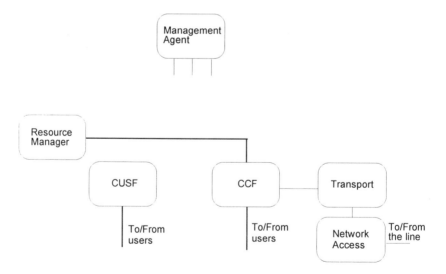

**Figure 2.5.7**
Overall functional architecture of the Italtel switching system

The Management Agent is based on SNMP and provides external control to the switching system. Transport and Network Access blocks implement the protocol stacks on which UNI and NNI signaling protocols are based. CCF and SSF respectively provide signaling and IN handling capabilities. Out-of-call events, like User-Service Interaction, are handled by CUSF. Finally, the Resource Manager block is responsible for the control of the ATM switching platform.

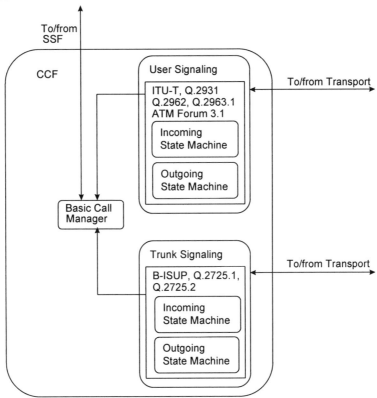

**Figure 2.5.8**
Functional organization of the CCF

Signaling protocol functionality is provided by user and trunk signaling, respectively, at UNI and NNI interfaces. The Basic Call Manager performs call control functions and is connected to the SSF in order to handle IN events.

## 2.5.2.2 Siemens Platform

The Siemens ATM node MainStreetXpress V2.1 was taken as platform for the B-SSP implemented by Siemens [2, 6, 7].

There are two configurations of the ATM switching system MainStreetXpress V2.1, one with a maximum of 62 ports, the other with a maximum of 16 ports. A port can be configured either as Subscriber Line Module Broadband (SLMB) or as Trunk Module Broadband (TMB) respectively serving the UNI or the NNI interface.

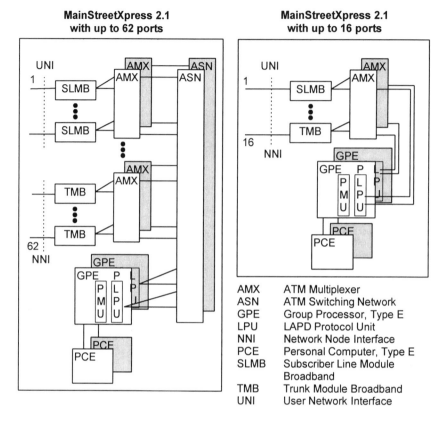

**Figure 2.5.9**
Hardware structure of the MainStreetXpress V2.1 switching system

These ports are connected to an ATM multiplexer (AMX). An ATM switching network (ASN) is used in order to interconnect these multiplexers in case of the large configuration. All the ports and multiplexers are controlled by the group processor GPE.

The GPE comprises the LAPD protocol unit (LPU) and the processing and memory unit (PMU). The LPU is in charge of the internal signaling, by which the ports and the multiplexers (AMX/ASN) are controlled. Additionally the LPU terminates the external signaling channels to customer premises equipment and to adjacent network nodes. The service layer of the UNI and the NNI is handled by the PMU. The GPE itself is controlled and administrated by means of the PCE. The PCE provides a user interface by which an operator has access to the whole system. By that, the PCE acts as a front-end without any stringent real-time requirements whereas the GPE realizes the time critical processing.

All hardware units, except for the ports, are doubled so that in case of a failure of a unit, a switch over to the assigned partner one can be performed. Therefore, a high degree of reliability can be achieved as is required by a public network.

In the following, some signaling characteristics of the MainStreetXpress V2.1

are pointed out.

At the UNI, switched connections and related signaling are restricted to VPI 0. Only point-to-point access configurations using a signaling virtual channel with VCI 5 are supported. The UNI protocol used (according to ATM Forum specification V3.1 or according to ITU-T recommendation Q.2931) can be defined by software configuration. The addressing mechanism provided is in accordance with the ITU-T recommendation E.164.

At the NNI, signaling is virtual path associated, i.e. a user channel set up via signaling has to belong to the same virtual path of the signaling virtual channel. The virtual channel with VCI 5 is always used as signaling channel.

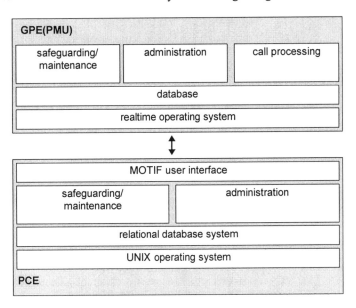

**Figure 2.5.10**
Software structure of the MainStreetXpress V2.1

In order to integrate the B-SSP functionality into the MainStreetXpress V2.1 the software running on the GPE and the PCE had to be modified. This software of the MainStreetXpress V2.1 is structured as shown in Figure 2.5.10.

The PCE runs a UNIX operating system. A relational database system is used in order to manage the persistent data of the application specific software of the PCE. This commonly available standard software provides the basis for the application software that is developed especially for this switching system. This special software can be subdivided into two blocks, one related to administration and one dealing with safeguarding and maintenance. The administration block includes functions for managing permanent connections and for subscribers of switched connections. The maintenance and safeguarding block deals with the startup and the status control of the system. For example, it performs recovery actions after detection of an error. Access to these application functions is provided to an operator by means of an OSF/MOTIF user interface. The application software is implemented using the C programming language.

Since the GPE realizes time critical operation and control functions, a real-time operating system is used. There is also a database to store the operating data of the switching system and to offer specific data access functions to the application software of the GPE. This application software is structured into three complexes. Administration and safeguarding/maintenance blocks are the back-end of the corresponding PCE complexes. Call processing includes functions for handling the upper layers of UNI and NNI protocols, resource management and through-connection of user channels (see [2, 6]). Using the IN terminology, the call processing complex represents the call control function (CCF). Both the database and the application software of the GPE are designed according to the object-oriented method. Consequently, an object-oriented programming language, namely Object-CHILL [8], is used to implement this software. Object-CHILL is an object-oriented extension of CHILL (cf. ITU-T Z.200 [9]).

## 2.5.2.3 Telefónica Platform

The RECIBA (Experimental Integrated Broadband Communications Network) project was developed by Telefónica in several phases starting from 1989 (see [10, 11]). RECIBA was also the core of the ATM backbone network necessary to carry out several international projects (ACTS, RACE,...) as well as the Spanish National Host at present.

The network nodes developed in the RECIBA framework were designed to build up flexible ATM VP/VC cross-connects or VP/VC switches with a number of different interfaces ranging from 2.048 Mbit/s to 622.08 Mbit/s.

VP/VC cross-connects use semi-permanent virtual connections which are established through OAM while VP/VC switches use signaling to establish virtual circuits. Both functions are supported by the RECIBA nodes depending on the software configuration of the equipment.

***RECIBA hardware overview***

The RECIBA node consists of a number of concentrators with up to 10 slots allocating different types of boards. It is composed of three subsystems (switching, concentrator and control) whose brief description is given in Figure 2.5.11.

- *Switching Subsystem.* RECIBA implements a *hyperlattice* topology where each node has three 640 Mbit/s links. One of these is attached to an access concentrator (I2 interface) while the others (I1 interfaces) are used to interconnect with other shelves. Each shelf contains a special board, called EBC, which is used to interconnect up to 64 shelves;

- *Concentration Subsystem.* This subsystem is attached to the I2 interface (proprietary 640 Mbit/s bus) . By means of request/response mechanism, it multiplexes the ATM cell flow coming from line interface boards (ITE). Each of these line interface boards contains part of the concentration logic as well as the master bus. They can also be configured to perform UNI or NNI interfaces;

- *Control Subsystem.* This subsystem governs the overall RECIBA equipment. It is implemented by the ECA boards and the workstation, attached via Ethernet. It also includes all the high level processes (e.g. the man-machine interface for operation and maintenance) and supports the signaling processes in the case of VC switch configuration. This subsystem collects and processes the OAM information from the ITEs and the EBC. The I2 interface (ATM bus) multiplexes the users' cell flows coming from the different ITEs and the control and management flows between the ECA and the EBC/ITEs . The EBC switches this multiplexed flow, routing the VPIs or VPI/VCIs towards and external I1 interface or the same I2 interface.

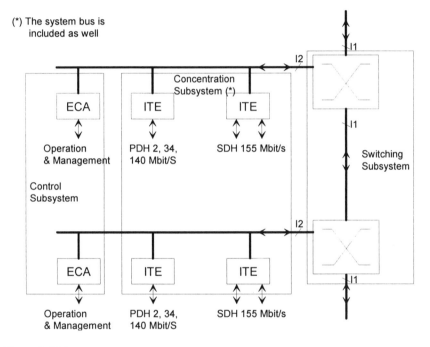

**Figure 2.5.11**
Hardware structure of the RECIBA switching system

### *RECIBA software overview*

The RECIBA platform has been designed following a distributed processing approach. All the boards have local processing capability.

Three software environments can be distinguished:

- The ITEs and EBCs running a Telefónica proprietary real-time operating system. The applications are focused in the low level management. Metasignaling functions are also present in the $S_B$ interfaces;
- The ECA running a POSIX embedded OS-9 operating system. It has functions to manage the concentration subsystem and it is also the gateway

between the high level functions, placed in the control workstation, and the ITEs. In case of VP/VC switches, it also supports the low level signaling protocols (SAAL) and the associated functionality;

- The control workstation, running under UNIX, provides the high levels of OAM and the graphical user interface. In case of VP/VC switches, it also supports the high level signaling protocols and the associated functionality.

### *Functional Architecture of the Switching Software*

Figure 2.5.12 illustrates the general software architecture of the RECIBA Switch.

RECIBA offers a friendly graphical user interface for OAM functions. This is supported in a SPARC workstation which sends/receives commands to/from the Resource Management and Routing (RMR) block.

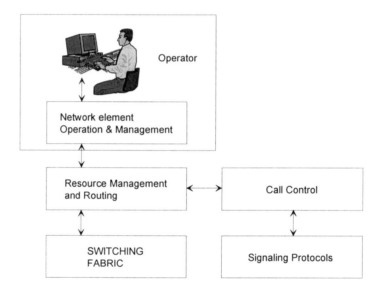

**Figure 2.5.12**
Software architecture of the RECIBA switching system

The main functions supported by the RECIBA Network Element OAM are:
- Configuration: Set-up, release and bandwidth modification of the semi-permanent VP/VCs;
- Fault: Alarms detection and fault identification;
- Performances: Policing function (based on leaky bucket algorithm), traffic monitoring, physical layer statistics and VP/VC connections trace;
- Charging: Pre-charging information based on the statistics of the call established by the users. This functionality is only available in the VP/VC switches;
- Security: Operator authentication functions.

The RMR block is responsible for the general control, reservation and activation of the Switching Fabric resources.

The overall control of the bandwidth is accomplished by the RMR. Whenever a bandwidth demand arrives, the RMR function checks the availability of the requested amount. In case of successful attempt, that amount is allocated and removed from the 'bandwidth pool'. If there is not enough bandwidth left, a rejection is launched.

Another functionality is the collection and primary processing of statistics, fault information etc. as well as the query on the status of some hardware elements.

It also performs the routing functions on demand of the Call Control (CC) entity. The RMR contains sub-blocks that are located close to the hardware and therefore they control all the hardware devices in order to manipulate cross connection capabilities.

Call Control carries out the overall control processing. At the same time, it acts as a link between the Signaling Protocol processes and the RMR.

The signaling implementation has evolved from the beginning of the RECIBA to the current INSIGNIA B-SSP, aligned with ITU-T UNI CS-2.1, with the following protocols:

- SAAL: I.363 (AAL-5) + Q.2110 + Q.2130;
- UNI: Basic call (Q.2931) and Q.2932.1 functionality (generic functional protocol: core functions) and compatibility with ATM Forum UNI 3.1 and point-to-multipoint call (Q.2971);
- MTP-3b: basically, contains the Signaling Message Handling functionality. The Signaling Network Management functionality is not included for simplicity (ITU-T Recommendations Q.701-Q.704);
- SCCP layer: provides only SCCP Connectionless Control (SCLC) function, class 0 (Basic connectionless class) and class 1 (In-sequence delivery connectionless class), with no Global Title Translation, and reduced management capabilities (ITU-T Recommendations Q.711-Q-713);
- TCAP layer: gives services to perform remote operations in a limited subset of functionality. Thus, only structured dialogs are allowed, the Dialog Portion of the TCAP message is absent and just Class 2 operations are used (ITU-T Recommendations Q.771-Q-775);
- B-ISUP layer: it is based on the ITU-T Recommendations (Q.2761-Q2764), with some restrictions: no interworking with N-ISDN is considered, only originating and terminating exchanges are supported, no international gateway functionality is provided, a subset of the maintenance procedures are implemented (VPCI reset).

## 2.5.3   Embedding a Service Switching Function into Signaling Software

The development of the CCF and SSF functional blocks caused significantly different kinds of software adaptation within the three platforms which were extended to B-SSPs within the INSIGNIA project. The CCF mainly comprises the usual B-ISDN signaling, which was already implemented in most of the platforms. Only the introduction of the feature of an SCP-initiated call and the

User-Service Interaction was new for the CCF of each platform. Other small adaptations of the CCF were necessary to provide the CCF/SSF interface.

In contrast, the SSF had to be implemented nearly from scratch for every platform. Since the SSF implementation of each B-SSP was object oriented, a big amount of common SSF development was possible for the different platforms, eased by the common models presented above in Section 2.5.1.

## 2.5.3.1 Case Study: Siemens B-SSP

A first example of reuse is in the design of the Siemens B-SSP. The software structure of the Control Processor (GPE) depicted in Figure 2.5.10 is refined in Figure 2.5.13, highlighting the main new part, the SSF, and its interfaces to the other software components. The hatched parts of these interfacing components indicate that only small modifications were necessary to integrate the SSF into the existing software (see [2]).

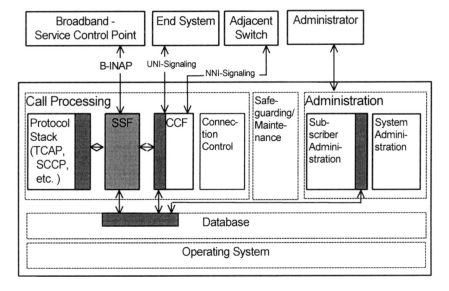

**Figure 2.5.13**
Software structure of the control processor (GPE/PMU) of the Siemens B-SSP

Figure 2.5.14 shows the object model for the design of the SSF. As it can be easily seen, the classes of the IN-SSM are present within the design (light gray boxes) with two exceptions. On the one hand, there are two classes for Bearer Connections. In theory, a Bearer Connection superclass could have been introduced as subclass of the SSM Connection class. However, since the Classes Bearer Connection and Connection coincided within the design of the first implementation, it was not useful to introduce a new abstract class without any meaning for the implementation.

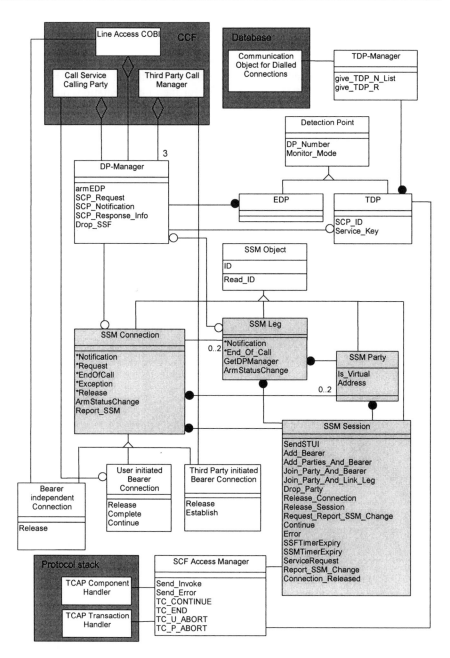

**Figure 2.5.14**
Design of the SSF for the Siemens B-SSP

On the other hand, only one leg class is present and no distinction between legs for bearer and bearer unrelated connection is made. Again, there was no reason for the implementation of such a distinction.

These examples show that a deviation from the analysis model is possible during the design, but must be well explained.

However, the provision of two specialized Bearer Connection classes (a class for user-initiated calls and one for SCP-initiated calls) instead of one general Bearer Connection class is useful for the implementation: both kinds of calls are realized differently within the CCF, but, from the viewpoint of a session, there are no conceptual differences. The same applies for the Bearer Independent Connection, which is again implemented independently within the CCF. The common superclass SSM Connections provides a uniform view of a connection to the Session class hiding the implementation details. This is a good example for information hiding by using polymorphism.

Furthermore, some classes are added to provide the functionality of functional blocks of the SSF: An *SCP Access Manager* object provides the functionality of the 'SCP Access Manager' functional block for a session, and the *object DP-Manager* performs the task of the 'Basic Call Manager' functional block for a specific basic call. The interfaces to other software components (dark gray boxes), as shown in Figure 2.5.13, are refined to associations between objects within the different software components. The relationship between SCF and SSF is not shown anymore, because it is realized implicitly by the communication over the protocol stack. However, the INAP-Messages, which are exchanged between B-SCP and B-SSP, appear as methods of the Session class.

Finally, there are three DP-managers assigned to one Third Party Call (i.e. one instance of the Third Party Call Manager). The reason for this is to model the independent views (on each of the two legs and the whole connection, see Chapter 2.2) necessary to handle a Third Party Call. Each DP-Manager is responsible for one of these views, communicating either with an SSM Connection object or an SSM Leg object.

The object-oriented approach was not only used successfully for the high level design, but it also provided powerful means for the low level design and the coding. For the Siemens B-SSP, a significant part of the existing software could be reused. Examples of reuse are the software development process and tools, the software architecture to structure the software, the design patterns for the communication of controller objects, the generic classes and, finally, subclassing to modify existing objects according to the new requirements. Some statistics regarding code reuse during the development of the Siemens B-SSP can be found in [12]. Earlier experiences with reuse at Siemens are documented in [13, 14].

## 2.5.3.2 Case Study: Italtel B-SSP

In the Italtel B-SSP, the adoption of object-orientation was particularly useful when the problem of introducing the IN functionality was addressed. Within the Basic Call Manager, an Interface class was present which was specialized for UNI and NNI interfaces. This class had a further specialization for IN, so the addition of the communication with the SSF was straightforward. Within the SSF itself, code reuse was successfully applied to deriving subclasses for the second experimental phase, where a distinction between bearer related and bearer unrelated connections and legs of the SSM was made. Figure 2.5.15 shows examples of reuse in the Italtel B-SSP.

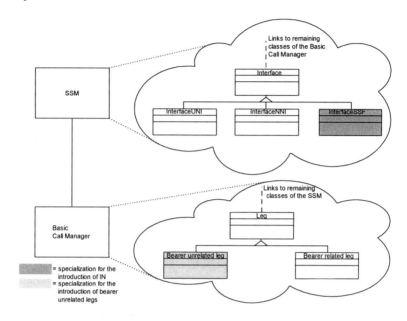

**Figure 2.5.15**
Examples of code reuse in the Italtel B-SSP

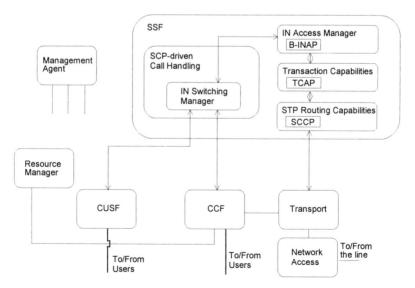

**Figure 2.5.16**
Functional organization of the SSF

The design of the SSF partitioned the problem in several sub-functions. Figure 2.5.16 shows the sub-functions of the SSF as well as the way that the SSF is integrated into the overall software architecture of the switching system.

The IN Switching Manager implements the INSIGNIA SSM and handles the resources used by the IN services. The remaining blocks in the SSF implement the necessary protocols in order to access a B-SCP.

The IN Switching Manager and the Basic Call Manager have been implemented in C++ while the remaining functions of theItaltel B-SSP are based on the C language. Implementations of B-INAP and TCAP protocols, specified in the ASN.1 notation, made use of tools for automatic code generation.

## 2.5.3.3 Case Study: Telefónica B-SSP

In the RECIBA B-SSP, the code evolution has been achieved by reusing the existing code. In order to keep the original RECIBA approach, formal techniques such as formal specification languages and the Object Modeling Technique (OMT) were adopted. Figure 2.5.17 shows this approach in the framework of the INSIGNIA project. All these changes were easier to be developed because of the use of these unequivocal procedures. On the other hand, the use of formal languages (SDL) and automatic code generators (SDT) considerably simplified the work of the developers and analysts.

The upgrade of RECIBA node to incorporate the B-SSP functionality and their main elements is depicted in Figure 2.5.18. The Call Control Function (CCF) and the Service Switching Function (SSF) are the principal components where blocks represent subsystems or groups of classes that provide a specific functionality.

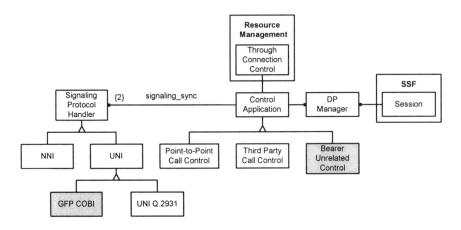

**Figure 2.5.17**
Design for reuse in the Telefónica B-SSP

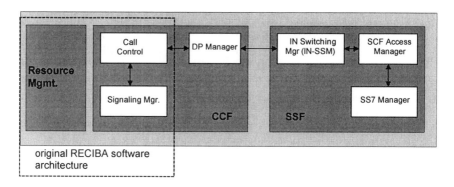

**Figure 2.5.18**
B-SSP architecture

## 2.5.4 Methodologies and Tools for Design and Development

A software engineering methodology is a process suitable for the organized analysis, design and implementation of software using a set of coordinated techniques. Basically three distinct methodologies have been used extensively in the framework of the INSIGNIA project: SDL (Specification Description Language), ASN.1 (Abstract Syntax Notation One) and OMT (Object Modeling Technique).

SDL and ASN.1, which are widely adopted in the context of telecommunication systems, have been used for realizing the INSIGNIA protocols, while the usage of the OMT methodology has been more focused on the application processes.

In particular, SDL helped designers in identifying the functional and dynamic behavior of the new protocols specified within the project, while the use of the ASN.1 language allowed a clear and consistent definition of the structures of the exchanged messages.

The OMT methodology is based on the modeling of the system under three different viewpoints; the threefold system model is subsequently refined and optimized in order to create the system project.

In OMT, the *Object Model* identifies the objects of the system and their relationships, the *Dynamic Model* describes the reactions of system objects to a given set of events and the interactions between objects, while the *Functional Model* specifies the way of modifying object parameter values and the constraints to these modifications. Systems designed using OMT methodology are more stable with respect to the specification changes and flexible in terms of code reusability, with respect to systems deployed using conventional approaches of software engineering (e.g. functional oriented).

Different tools for design and automatic generation of software which are based on the above mentioned methodologies have been adopted by designers and implementers. SDT™ (SDL as a Design Tool - by Telelogic) has been employed both as a graphical editor and automatic code generator for realizing some protocol layers. A public release of the ISODE (ISO Development

Environment - by ISODE Consortium) software has been used in order to realize procedures for the automatic encoding/decoding of messages exchanged between some protocols and/or application processes specified by means of the ASN.1 syntax. Finally, the adoption of an Object Oriented analysis and design tool called Paradigm Plus™ (by Platinum Technology), which provides automatic generation of Object Oriented code and advanced features like Reverse Engineering, has provided the designers with powerful support for the creation of flexible systems.

# References

[1]    Rumbaugh J et al., *Object-Oriented Modeling and Design*, Prentice Hall, 1991
[2]    von der Straten G and Totzke J, *An Object-Oriented Implementation of B-ISDN Signaling*, Proceedings of ICC'94, New Orleans, 1994
[3]    Marino G, Merli E, Pavesi M, Rigolie G, *A hardware platform for B-ISDN services multiplexing: design and performances of AAL and ATM layers*, Proceedings of GLOBECOM'93, Houston, 1993
[4]    Balboni G, Collivignarelli M, Licciardi L, Paglialunga A, Rigolio G, Zizza F, *From transport backbone to service platform: facing the broadband switch evolution*, Proceedings of ISS'95, Berlin, 1995
[5]    De Zen G et al., *A high level management system for an ATM-based Service Multiplexer*, Proceedings of the IEEE ATM workshop, San Francisco, 1996
[6]    Totzke J, Welscher J, *A Prototyped Implementation of B-ISDN Signalling at the Network Node Interface*, Proceedings of GLOBECOM'95, Singapore, 1995
[7]    Fischer W et al., *A Scalable ATM Switching System Architecture*, IEEE J. Select. Areas Commun., **9**, 1299-1307, 1991
[8]    Dießl G, Winkler J F H, *Object-CHILL - An Object-Oriented Language for Systems Implementation*, Proceedings of ACM Computer Science Conference '92, 1992
[9]    ITU-T Recommendation Z.200, *CCITT High Level Language (CHILL)*, 1988
[10]   Fernández-Vega L, Chas Alonso P L, *Proyecto RECIBA: Telefónica hacia la RDSIBA*, Comunicaciones de Telefónica I+D, December 1992
[11]   Lizcano P J, *A flexible architecture platform for B-ISDN interfaces and services benchmarking*, GLOBECOM'94, San Francisco, 1994
[12]   van der Vekens A, Urban J, Hussmann H, Fabrellas A, Venieris I, Zizza F, *Object-Oriented Realization of a Broadband Intelligent Network Architecture*, in Proceedings of ISS'97, Toronto, 1997
[13]   van der Vekens A, *An Object-Oriented Implementation of B-ISDN Signalling - Part 2: Extendibility Stands the Test*, Proceedings of 18th ICSE, International Conference on Software Engineering, Berlin, 1996
[14]   Guenther W, *Applying Object-Oriented Reuse Techniques to Switching System Software*, ITG Fachbericht 137, VDE Verlag, 187-194, 1996

# Chapter 2.6

# THE BROADBAND SERVICE CONTROL POINT - B-SCP

This chapter looks in detail at the role of the Service Control Point (SCP) integrated into a Broadband network. The SCP constitutes the heart of the IN network as it incorporates the main intelligence required for the provision of advanced multimedia services. An overview of the hardware and software architecture is given along with a discussion of the methodology used for the development of the service logic programs. Examples of the Video on Demand and Broadband Video Conference services, presented in Chapter 2.4, are given.

## 2.6.1   Introduction

There are two different ways to look at the role of a Broadband Service Control Point in an Intelligent Broadband Network: the service engineering point of view and the network evolution point of view.

From the service engineering point of view, the Broadband Service Control Point is a very specific implementation architecture which is not mandatory in general. The important issue for service engineering is to identify common and reusable functionality which can be applied to several Broadband multimedia services. Some of this reusable functionality is related to control of network resources which is a good reason to implement it on a physical entity which is closely related to network control, like the Service Control Point. However, it has to be kept in mind that in most cases a Broadband multimedia service also contains functionality which is not related to the network and which therefore is better implemented on the end systems or on special servers. In an Intelligent Broadband Network, the Broadband Intelligent Peripheral (B-IP) is devoted to play the role of such a specific server for services.

From the network evolution point of view, it is important to provide for a smooth transition from narrowband to Broadband public networks. Therefore the details concerning this transition should be transparent to the Service Creators as well as the Operators. Fortunately, as discussed in Chapter 2.1, the role of an SCF in a Broadband environment does not change significantly from its use in the narrowband context. There is a difference in the type of services to be controlled by the B-SCP but the internal architecture of a B-SCP is more or less identical to that of a narrowband SCP. Another difference is that in narrowband the service is always related to a single call, whereas for Broadband the service will relate to a session which may comprise a complex collection of calls. A

*Intelligent Broadband Networks.* Edited by I. S. Venieris, H. Hussmann
© 1998 John Wiley & Sons Ltd.

B-ISDN network has potentially the capability of supporting several parties and connections to form a single session. However, the main part of session management can be achieved by the B-SSP, so that there is only little impact on the B-SCP architecture from this side, too. Functionally, the Broadband SCP will contain the SCF and may contain an SDF. The B-SCP is connected to B-SSPs and optionally to B-IPs through the signaling network. The B-SCP can also be connected to a B-IP via a B-SSP relay function. These connection options are not different from those possible in narrowband.

The concept of a Broadband Service Control Point is an interesting case of reuse of architecture. There are several alternative architectures for implementation of multimedia services, in particular approaches which completely rely on generic software communication architectures like CORBA. However, the Broadband Intelligent Network has the advantage of reusing all the experience achieved with narrowband SCPs and also inherits the complete tool set (Service Creation Environment) which was developed for the narrowband context.

## 2.6.2   Design Methodology for Service Logic Program Development

The Service Control Point (SCP) is the key network element in an IN based environment since it is the point where the execution of the major functions of the Service Logic Programs (SLPs) takes place. The purpose of the process of service logic program development is to provide such an SCP with the appropriate software which will realise a Broadband IN telecommunication service.

The successful outcome of the SLP development process requires the co-operation between various categories of people and systems. In terms of people, there are users, analysts and developers involved. In terms of systems, there are the IN elements and the tools used for specification and development of a service. Each one of these factors makes its own impact on the quality of the delivered service. User needs and preferences should be taken into account since telecommunication services are expected to be commercially appealing. Analysts should attempt to combine these needs with the available network infrastructure as well as with the existing development tools and produce a service specification. Developers use this specification as a guide for writing the service logic program. A well defined design methodology undertakes the role of combining these factors in order to guarantee the delivery of Broadband multimedia services.

The proposed methodology is presented here in the form of discrete phases in Figure 2.6.1. The starting point is a prose description of the service from the user's point of view. This description is used for determining what part of the functionality of the service will be assigned to each network element. Consequently the interfaces in the Distributed Functional Plane (DFP) between these elements are specified. This information is then aggregated into a formal description of the service. Formal description is used as a guide for authoring service logic code by means of a Service Creation and Simulation Environment.

Finally the service, after being successfully tested and tuned, becomes operational.

Each development phase is explained in detail by means of two practical IN service examples: a Video on Demand (VoD) and a Broadband Video Conference (B-VC) service. Both services have been introduced in Chapter 2.4 above. VoD [2] is a service which enables a home user to select movies and television programs and view them on a TV set equipped with a set-top box. Moreover the user has full VCR functionality like rewind, pause etc. The B-VC service [3] enables groups of users to set-up virtual conferences where each of the conferees can participate from his/her own premises.

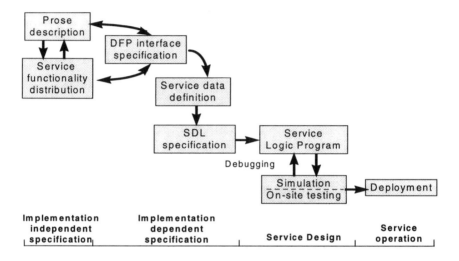

**Figure 2.6.1**
Specification and design methodology

Prose description is the initial stage of the development process; it is only natural to begin with the user preferences. Each service is decomposed in discrete functional stages. This decomposition facilitates the description itself as well as the transition onto the next development phase. For the VoD service such a decomposition would result in Figure 2.6.2. For each functional stage, a text scenario is provided which is based on the fact that the service prompts the user, the user responds and the service logic, based on that response, decides what to do next.

A graphical representation for each scenario is given using service independent building blocks. This is a kind of initial flowchart that will be evolved in a formal description during subsequent development phases. For example the Service Provider Selection stage of the VoD service is described in the manner shown in Figure 2.6.3. Implementation specific details are not examined at this point and the only assumption made is that the user possesses a multimedia terminal of some kind in order to interact with the service logic.

Once an IN service has been initially specified by the way described in the previous paragraphs, the exact responsibilities of each IN node should be defined. In the classical view on service logic program development, the service logic is executed entirely within the B-SCP, switching functionality is implemented within the B-SSP, the SDP is in charge of database transactions and the B-IP provides the resources needed (messages, menus etc.) for communication with the CPE.

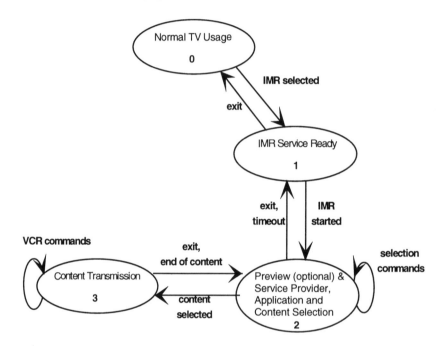

**Figure 2.6.2**
Decomposition of VoD service

The software running inside the CPE is usually an application with a graphical user interface combined with a UNI signaling stack. In real-life implementations, the SDP is combined with the SCP in the same physical entity. In the case of narrowband services, the SCP may also incorporate the B-IP.

In the case where Broadband services are concerned, the B-IP should be separate since it has to support multimedia resources which are much more demanding. Putting the B-IP in a separate machine results in the fortunate side effect of having processing power to spare (e.g. in the execution of service logic functions). Therefore, instead of placing all of the service logic into the Control Point we can split it by keeping a main controlling entity in the Control Point and moving the discrete functional stages of our service in the Intelligent Peripheral in the form of autonomous modules. This technique is also aligned with the ITU-T specifications for CS-2 [4] where the role of 'scripts' (integrated medium-sized programs executing IN service functions) is investigated. Also the data

used by the Service Logic have to be divided between the Control Point and the
Intelligent Peripheral. The resulting configuration is presented in Figure 2.6.4.

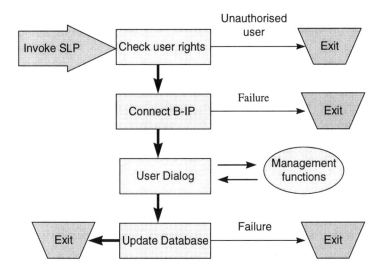

**Figure 2.6.3**
Decomposition of service provider selection

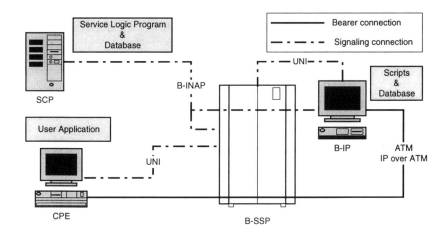

**Figure 2.6.4**
IN functionality distribution

Another advantage of the IN concept is that the communication between the
CPE and the B-IP can be achieved by using a variety of protocols. The obvious
choice is a direct ATM bearer connection but in case this is not the best solution

other alternatives exist like Internet Protocol over ATM. Using Internet Protocol over ATM presents two major advantages to the potential developer: (a) Porting applications originating from other environments like Ethernet into the IN concept with minimal changes (b) Use of platform-independent software (e.g. HTML - Java) for the creation of the interfaces handling user to service interaction. In the examples presented in this section, the scripts approach has been used for the VoD service while the classical approach has been used for the B-VC service.

Having decided on the issue of service functionality distribution and before producing a formal service description, the interfaces between the various network elements must be specified as well as the data that will be used by the service logic program. Current experience shows that the only layer of the IN protocol stack that requires modification when moving from narrowband to Broadband networks is the INAP layer which controls the communication between the switching and the control point. The Broadband INAP (B-INAP) layer presented in this book has been designed having in mind the fact that it should be able to cater for new Broadband services without changes. If changes are required, it is preferred to add new B-INAP messages since this does not affect other already running services. However, in the general case of completely new services, it may be unavoidable also to modify some of the existing B-INAP messages. While this is true for the B-INAP presented herein, it should be strongly avoided in the case of a future ITU-T approved B-INAP.

At this stage of IN service specification, the service scenarios are depicted in the form of message sequence charts. Initially, these message sequence charts are used for identifying possible problems like the need for adding new IFs or eliminating redundant ones. When they are finalized, they are used as input for the formal specification phase. Examples of such message sequence charts can be found in Chapter 2.4 above. If changes to the B-INAP layer are required, they have to be formally specified in the abstract syntax notation (ASN.1), enhancing the existing ASN.1 specification.

The data handed by the service logic could be subdivided as: service data (for example fixed data and announcements), management and subscription data (for example subscription profiles), user data (for example user profiles) and call instance data (for example counts against unauthorized access attempts). The definition of service and call instance data can be decided during the formal specification phase. On the other hand, management, subscription and user data should have been defined in an earlier phase. An incomplete or erroneous definition of these data structures almost certainly results in modifications during the implementation phase therefore causing severe delays in service development and deployment. Regarding the location of the service data tables, each table is located at the same network node where the respective service logic function is executed.

The specification process is completed by combining the previous phases into a formal service description. Various methods can be exercised for formal description production, for example a flowcharting tool can be used. The method proposed in this book is the use of the Specification and Description Language

(SDL) [5] for a number of reasons. First of all, this language is an international standard created specially for the description of telecommunication systems.

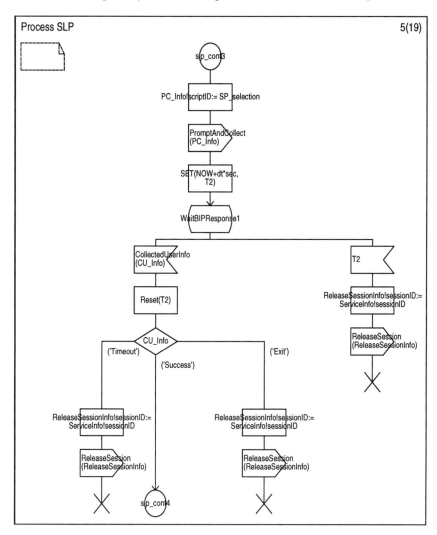

**Figure 2.6.5**
Description of the VoD service B-SCP in SDL

Also, some experimental service creation tools [6] include verification modules and SDL which makes a good candidate for use with such modules in a future commercial product. Moreover, in the rare case where a service creation environment is not available then SDL can be used for verification, simulation and automatic production of source code for a telecommunication service [7, 8].

A proposed representation of an IN service in SDL is based on the following set of rules: The IN service is represented by the SDL *system* structure. The network elements involved in service logic execution are represented by the SDL *block* structure. The connection between the various network elements is represented by SDL *channels*. Messages exchanged between IN network elements are represented by the SDL *signal* structure. Each service stage, described in the prose description, is represented by an SDL process. Each process is connected to the SDL environment via channels. Since the only channels defined are those connecting the IN network elements each process is receiving/sending signals via those channels. For the VoD service, the control point contains only one such process which controls the execution of the intelligent peripheral scripts. The SDL diagram for the Service Provider Selection part of the VoD service is presented in Figure 2.6.5. The diagram shows the logic behind the Service Provider Selection stage. Initially the SCP demands the execution of the appropriate script from the B-IP. Eventually either a reply will arrive or the operation will time-out. Depending on the outcome the service logic will follow the appropriate path.

The actual service logic creation process as well as the available tools are discussed in the following sections.

## 2.6.3    Platform Architectures for the Design and Execution of Service Logic

In this chapter, the specific features of the prototype B-SCP developed for the INSIGNIA project are discussed. However, it should be noted that whilst this B-SCP is a specific realization for a research project the design principles and infrastructure discussed are equally applicable for a B-SCP when deployed in a real B-ISDN network.

The essential items for INSIGNIA were the definition and realization of the Broadband communication stack and the design and application of the IN Switching State Manager. The former has impact on the B-SCP while the mechanism for the latter is covered within the SCE and embedded in the service logic programs when they are deployed on to the B-SCP.

### 2.6.3.1 Realization of the Prototype B-SCP

The prototype B-SCP is realized as design enhancements to an existing narrowband SCP from the GAIN product range produced by GPT LIMITED. The B-SCP utilizes the existing infrastructure of a narrowband SCP which provides a Service Logic Execution Environment (SLEE) and an internal relational database.

To accommodate the Broadband environment a new protocol stack has been integrated and the Broadband INAP message set has been designed into the B-SCP. In addition an interface to a B-IP has been included. Services are deployed to the B-SCP via a management interface and some management facilities such as logging are available. Features to handle performance and overload situations are also covered.

Service Logic Programs (SLPs) are created using an adaptation of the GPT Service Creation Environment (SCE) used for narrowband application. The narrowband SCE comprises a Service Definition System and a simulator for testing services.

These enhancements are described in further detail below.

## 2.6.3.2 Hardware Architecture

The presentation starts from a high level view of the system architecture for the prototype B-SCP. Figure 2.6.6 provides a physical block diagram of the architecture of the B-SCP. The physical architecture is based on a GPT SCP narrowband product called GAIN. It should be noted that the normal configuration of the narrowband SCP product comprises additional features in order to provide a robust, high performance platform. These features were removed from the prototype B-SCP, however, when the B-SCP is deployed in a real B-ISDN network the additional features would be required.

The B-SCP is composed of several workstations (from SUN Microsystems Inc.) which are interconnected over an Ethernet LAN. The SCP comprises workstations of two different types which are called the traffic Computing Element (CE) and the Local Manager (LM). Processor CE is dedicated to call handling and does not host any service management software. The prototype B-SCP was limited to one CE but the normal configuration for a B-SCP would support several CEs. The LM is the primary point for service management of the B-SCP features. The management software resides on this processor and is used by both local and remote management users. Interaction between the LM and the CE is achieved by passing messages via the Ethernet LAN.

On the LM, a modem is provided to permit access by customer support staff. Several tape drives are used for system and adjunct software archiving, for continuous backup of the database transaction log and for service archiving and restoration. A printer provides report and error logs from the service management.

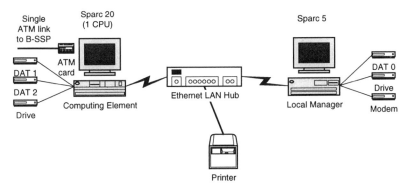

**Figure 2.6.6**
SCP Physical Architecture

## 2.6.3.3 Software Architecture

### Local Manager (LM) Software Architecture

The software architecture of the B-SCP local manager (LM) is shown in Figure 2.6.7. The management user interacts with the Service Management Environment (SME) which provides service management views and access to the UNIX views for platform management. The platform management provides capabilities to perform functions such as system archive/restore and data transfers, monitor for alarms, overload, statistics etc.

The Simple User Interface (SUI) provides a command line interface to gain access to managed objects which cannot be accessed via the SME. The Log SUI provides a command line interface to gain access to the SLEE logs. The SME has the capability to access multiple SLEEs on the CEs via the SLAC (SLEE Access Controller) but in the case of the prototype B-SCP only one CE is present, the SLAC is therefore redundant. DCOMMS provides a facility for passing messages between the SLAC and the remote SLEE over an Ethernet link.

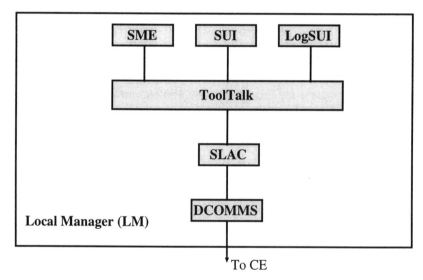

**Figure 2.6.7**
LM software block diagram

### Computing Element (CE) Software Architecture

Figure 2.6.8 is a diagram of the CE software structure, for reasons of clarity it includes two logical instances of the relational database. The logging data instance is used for SLEE logs, while the service data held in the database is accessed via the SLPDB when running the various SLPs.

The Switch Control (SWC) for the B-SCP interfaces to the SLEE and to the B-SSP. At the component level, SWC converts ASN.1 encoded messages from the TCAP router into SLEE block syntax for the set of B-INAP operations and routes the message to the Service Logic Programs (SLPs).

The ATM Stack Handler is responsible for handling the NNI signalling interface from the physical layer to the SWC.

The SLEE provides an environment for the concurrent execution of different SLPs in order to deliver multiple services. It is accessed by the SLPs through a standardized (Bellcore AIN) SLEE API which shields the SLPs from implementation details. The SLEE API comprises a set of service independent functional components (FC). The FCs enable the SLPs to use local resources and through message communications with SWC to command the SSP call processing. Behind the FCs there is a collection of SLEE processes that divide into subsystems for measurements, logging, SLP scheduling, timing and system management.

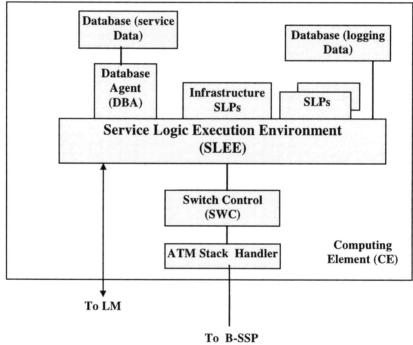

**Figure 2.6.8**
CE software block diagram

Having presented the basic infrastructure of a B-SCP in terms of hardware and software modules, it would be interesting to observe the exact way by which the SSM state model is mapped inside the B-SCP.

## 2.6.3.4 Broadband Service Creation Environment

The Broadband Service Creation Environment (B-SCE), described here, is based on an existing GPT product, GAIN INventor [9]. INventor is an object-oriented graphical service creation environment (GSCE) which enables designers of telecommunication services and marketing staff to build and simulate services from scratch. A key feature of this SCE is that it allows a user to create services rapidly with no knowledge of programming. In addition, it isolates the user from the underlying IN implementation, and allows the user to express a service in a sufficiently abstracted way.

To support the creation of advanced Broadband multimedia services, the service creation environment has been extended to cater for the abstract view of the enhanced SSM. The changes to the GAIN INventor were made to take into account the new B-INAP, which resulted in new Generic Service Building Blocks (GSBBs) being created. The GSBBs are the GPT's approach to the Service Independent Building Block concepts of IN CS-1 [10] but provide a finer grain of control than the latter. The GSBBs are generic and tailored to specific interfaces and their associated procedures. This approach allows the creation of new services without the need to redevelop the software which comprises the basic service creation infrastructure. Since the GSBBs are tailored to the interface specifications, changes to peripherals can be more easily accommodated. An Icon is the visual representation of a GSBB within the sketcher tool of the GSCE.

The narrowband SLEE has been reused with some slight modification to introduce the SSM model view at the service logic infrastructure. The B-INAP independent GSBBs were re-used without any changes. The newly created GSBBs allow the manipulation of connections and parties and the interaction with the B-IP.

The approach within B-SCE has been to hide the complex view of the objects and relationships from the service creator. Instead of providing a complete view of the SSM, the service creator is only aware of parties and bearer related/bearer unrelated connections. The service creator uses the directory number to reference the party objects and uses the identifiers to identify bearers. It has no actual visibility of leg objects. The service infrastructure within the B-SCE handles the complex association of objects and identifiers performing the necessary mapping onto the enhanced SSM model.

## 2.6.4   Intelligent Network-based  Service Examples

The GSBBs which have been implemented for the Broadband Service Creation Environment are organized in an Icon library that is used for the development of Broadband IN service pictures. The library of Icons is presented to the service creator as a palette, as depicted in Figure 2.6.9. The service creator can then pick and drop from the palette to create the service picture. The Icon library is the medium by which the developer can transform the SDL specification of the IN service to the actual service logic program.

The Broadband Icon library is a mixture of existing narrowband Icons (some of which have undergone slight modifications) and entirely new Broadband ones. Special attention should be paid to the following Icons which are directly related to Broadband functions:

*Play*: This Icon is used for the communication between the user and the service via the B-IP. A PromptAndCollect information flow is sent to the B-IP which prompts the user (either directly or by executing the appropriate script), collects the user answer and returns it to the SCP.

*IPprompt*: This Icon is a variant of the Play Icon. A PromptAndCollect information flow is sent to the B-IP which prompts the user (either directly or by executing the appropriate script). This information flow may optionally include extensions to the service related data.

*IPmonitor*: This Icon is a variant of the Play and Prompt Icons. It is used to monitor for (a) responses from the user as a result of a PromptAndCollect information flow which may contain extensions to the service related data and (b) any switch related events.

*Party*: This Icon is used for party related operations like adding new parties or dropping existing parties.

*Bearer*: This Icon is used for either adding or releasing a bearer connection between two parties of an IN service.

*Join*: This Icon is used to route user initiated calls to introduce a new party into a session.

**Figure 2.6.9**
Broadband Icon library.

Also a special kind of Icon called the Nested Icon is used for grouping Icons performing a specific operation.

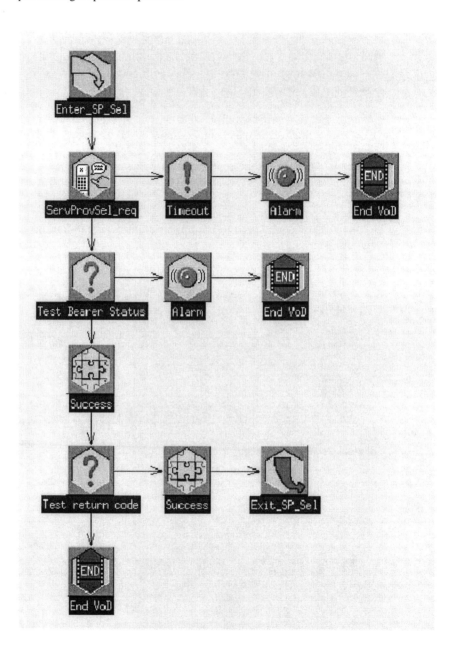

**Figure 2.6.10**
VoD Service Picture.

The Service Provider Selection stage of the VoD service, which is depicted in Figure 2.6.10, demonstrates the exact way by which the Icon Library is used for sketching a service logic program picture. The Enter_SP_Sel Icon is the entry point. The service logic attempts to setup a bearer connection between the User and the B-IP. If the connection is successful, it demands the execution of the Service Provider selection script and collects the result. This is achieved with the ServProvSel_req Icon. If a result never arrives from the B-IP then the Timeout Icon is activated. Consequently the Alarm Icon is used for registering an appropriate alarm message in the alarm view of the platform. In this error case, the session is released with the End VoD Icon. The Test Bearer Status Icon sets the Bearer Status variable as the variable that will be checked compared against the 'Success' result with Icon. If the bearer setup is successful the service logic passes from the Success Icon. In case of failure, the Alarm Icon registers an appropriate alarm message in the alarm view of the platform and the End VoD Icon terminates the session.

In the case of bearer setup being successful, the service logic moves to examine the result returned from the B-IP so it sets the Return_Code variable as the variable that will be checked compared against the 'Success' result with Icon. If the service provider selection is successful, the service logic exits through Icon Exit_Sp_Sel. Otherwise, an error has occurred and the session terminates with the End Vod Icon.

The B-VC is a more complex service since it should correlate a large number of connections and offer features to the service user such as, create conference, establish conference, manage conference (either dynamic or static), add user, join conferee, disconnect user etc. Chapter 2.4 above gives more detailed information on the features of the B-VC service. As an example of the B-VC service, an abstraction of the create conference interactions is depicted in Figure 2.6.11.

Figure 2.6.11 provides an overview of the conference creation stage of the B-VC service. The service is invoked on detection of a unique IN number. Service start-up procedures consist of the initialization of various variables and flags. This part of the service can be realized as a separate SLP. These start-up procedures have not been included in Figure 2.6.11.

On successful completion of the start-up procedures, the service enters the conference creation stage. The calling user is initially connected to the B-IP. A test is performed to determine whether the IP connection was successful. Failure would result in the termination of the session. The B-IP is then requested to obtain the conference details from the calling user and to return this to the SCP for storing in the database as the Conference Information Record (CIR). These actions are illustrated in Figure 2.6.11 by the Icons IPprompt, IPmonitor, IPResponse and Database. For reasons of simplicity, only one occurrence of the IPprompt, IPmonitor and IPResponse Icons are shown, but depending on the nature and sequence of the information being requested from the calling user, these actions may be repeated several times. Also, only a minimum of error checking is shown, for example, checking the user response for different conditions following the IP interaction has been omitted.

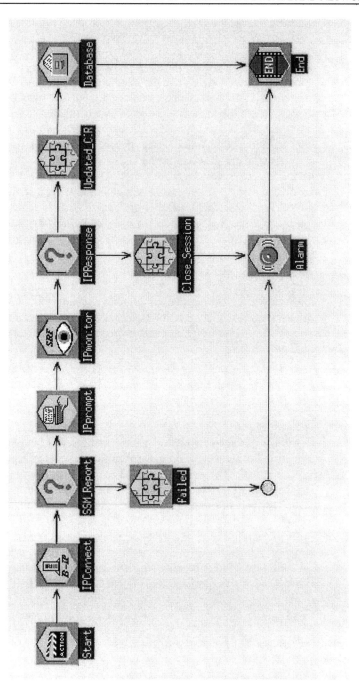

**Figure 2.6.11**
B-VC Service Picture.

On successful completion of the conference data entry by the calling user, the Database Icon performs the necessary database operations for the creation of a new conference record. Checks for failure conditions following the database update have not been included in Figure 2.6.11.

Finally either a successful or an unsuccessful (for any reason) termination of the Conference Creation SLP is achieved with Icon End.

# References

[1] Prezerakos G N, Venieris I S, *Specification and Design of Interactive Multimedia Services for Broadband Intelligent Networks*, Intelligent Networks and Intelligence in Networks (Edited by Dominique Gaiti), Chapman & Hall, Proceedings of IFIP TC6 WG6.7 International Conference on Intelligent Networks and Intelligence in Networks (2IN97), Paris, France, 1997

[2] Vezzoli L, Lorenzini L, *Intelligent Mobile Video on Demand*, Proceedings of ICIN '96 Conference, Bordeaux, France, 1996

[3] Brandt H, Tittel C, Todorova P, *Broadband Video Conference in an Integrated IN / B-ISDN Architecture*, Proceedings of NOC'97, Antwerp, Belgium, 1997

[4] ITU-T Recommendation Q.1228, *Interface Recommendation for Intelligent Network Capability Set 2*, 1997

[5] ITU-T Recommendation Z.100, *SDL Methodology Guidelines*, Appendices I and II, 1993

[6] Sakai H, Takami K, Niitsu Y, *Service Logic Program Generation Method using Specification Supplementing Techniques*, Proceedings of IN'97 IEEE Workshop, Colorado Springs, May 1997

[7] Nyeng A, Møller-Pedersen B, *Approaches to the specification of Intelligent Network Services in SDL-92*, in SDL'93: Using Objects, Elsevier Science Publishers, 427-440, 1993

[8] Olsen A, Nørbœk B, *Using SDL for Targeting Services to CORBA*, Proceedings of IS&N 95 Conference, Heraklion, Crete, 1995

[9] GPT LIMITED UK, *GAIN 300 Inventor User Guide for release 304.2*, 1996

[10] ITU-T Recommendation Q.1213, Global Functional Plane for Intelligent Network Capability Set 1, 1995

# Chapter 2.7

# THE BROADBAND INTELLIGENT PERIPHERAL - B-IP

## 2.7.1   Introduction

The Broadband Intelligent Peripheral (B-IP) is a multimedia agent supporting the communication between the user and the Service Logic. The provision and support of advanced complex multimedia services over an IN infrastructure requires a more advanced and significantly different type of Intelligent Peripheral than in the narrowband case. For example, as it was already explained in Chapter 2.2 above, a Broadband IP contains specific service logic which complements the service logic program executed on the B-SCP. In this chapter, models and architectures are presented for providing a B-IP which supports complex multimedia services. The philosophy of these models is guided by the need for modularity and expandability of the Peripherals in order to meet the needs of newly introduced services. Examples of an Interactive Multimedia Retrieval and a Broadband Video Conference service are presented.

## 2.7.2   Requirements

The goal is to provide a B-IP that is capable of handling not only services based on sophisticated user interactions deploying multimedia features, but also to be able to anticipate the introduction of new services in the future. Therefore, the B-IP model needs to be based on a modular architecture. In more detail, the B-IP architecture must allow for :
- The support of different protocol stacks for communication with the user;
- The quick and easy replacement or enhancement of the protocol stack used for communication with the SCP;
- The quick and easy introduction of service logic parts necessary for supporting new services.

An adequate architecture for these purposes should feature a core administrative entity using standard interfaces for interfacing and managing different modules (i.e. protocol stacks, service logic programs, devices).

Furthermore, the service logic parts executed on the B-IP must also be based on a flexible service design to allow for:
- The easy introduction of services and re-use of already existing service parts to form new services;

*Intelligent Broadband Networks.* Edited by I. S. Venieris, H. Hussmann
© 1998 John Wiley & Sons Ltd.

- The minimization of SRF-user interaction by using compact, self-contained service logic program parts that can serve integrated segments of user interaction.

For this reason, an object-oriented architecture for the provision of the service logic executable parts on the B-IP can be used which was inspired by the idea of 'User Interaction Scripts' [1]. The term *User Interaction Script (UIS)* [2] denotes a compact, self-contained set of instructions that upon execution can implement a functionally integrated part of a user interaction for supporting an IN service.

## 2.7.3   Broadband Intelligent Peripheral Architecture

### 2.7.3.1 Basic Architecture

In Figure 2.7.1, the physical and software architecture of a B-IP based on the requirements presented in the previous paragraphs is presented.

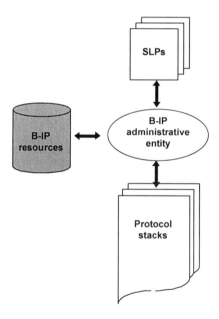

**Figure 2.7.1**
B-IP architecture

In this scheme we can distinguish four different entity types:

The *B-IP administrative entity*. This is the core of the B-IP and is used to trigger, interface and supervise the operation of all other B-IP modules. We can consider it to be the heart and mind of the B-IP since all actions performed on the B-IP have to be decided and initiated by this entity.

The *protocol stacks*. Here, we can distinguish two types of protocol stacks; those which are used for interfacing the user (both control and user data plane) and those which are used for interfacing the SCP. For the user related protocol stacks, a UNI protocol together with a transport protocol for transferring the data related to the user interaction are necessary, while for the interaction with the SCP an INAP or B-INAP protocol is necessary. All messages arriving in the B-IP which are related to service support SRF functionality are passing from the B-IP administrative entity which translates them to the responding actions. On the other hand, all requests and messages leaving the B-IP are forwarded to the related protocols by the B-IP. For this reason, a protocol interface module for each protocol stack is embedded in the B-IP administrative entity.

The *Service Logic Programs.* These are the service logic parts necessary for implementing the SRF functionality on the B-IP. Each IN service that deploys a B-IP has a corresponding SLP instance on the B-IP that upon execution provides the necessary logic for serving SRF-user interaction. To communicate with the B-IP administrative entity, a standard SLP interface is used for supporting the action requests from the service logic and for transferring the collected user information as a result of an SL execution.

The *B-IP resources.* These are the SRF resources deployed from the services supported by the B-IP. Such resources for a B-IP include speech recognition and speech synthesis devices, video encoders/decoders and others. These devices are used to facilitate a multimedia interaction between the SRF and the user.

## 2.7.3.2 Service Logic Implementation Architecture

Closely related to the architecture and implementation of a B-IP is the method adopted for the realization of service logic residing on the B-IP. Following the emphasis on modularity which was pointed out in the description of the B-IP architecture, a modular technique is adopted also for the provision of the service logic parts executed on the B-IP. Such an approach encompasses the ability for expandability and reusability of the existing service logic parts for future implementations of new services. The idea behind this modular service logic implementation is the use of User Interaction Scripts (UIS) capable of handling integrated parts of SRF-user interaction. This leads to a substantial decrease of network traffic over the SCF-SRF interface since only message exchanges related to the triggering and reporting of script execution actions are necessary. Furthermore, the adoption of UIS offers a connecting bridge between the service definition on the Service Plane and the Global Functional Plane [3] by providing the means for a high level description of service parts that can be reused for providing other services (i.e. authentication script, authorization script, user selection script). Finally, an object-oriented method for the description of these scripts will contribute to the modularity and extensibility since it will facilitate the process of modification or enhancement of already existing scripts according to requirements posed by new services (i.e. enhancement of an authorization script based on password checking, to support also speech recognition).

Following this UIS based description of the service logic parts implemented on the B-IP, we can describe a simplified model for SRF-SCF interaction based on message exchange related to the execution of scripts. For this reason, new messages have to be added to the basic B-INAP protocol between the SCF and the SRF as it was presented in Chapter 2.3 above.

The following Information Flows (IFs) are added:

- *Script Close:* This IF is issued by the SCF to de-allocate the resources used for invoking the execution of a script. It is used to terminate the sequence of actions for executing a script;
- *Script Event:* This IF is issued by the SRF to return information to the SCF on the results of the execution of the instance of a script. This result might be the partial result during the execution of the script or the final result after the execution has been completed;
- *Script Information:* This IF is issued by the SCF to send to the SRF additional information about a script execution;
- *Script Run:* This IF is issued by the SCF to allocate the necessary resources to create the instance of a script and then to activate this script instance. This script may be related to a User interaction or an internal B-IP management function.

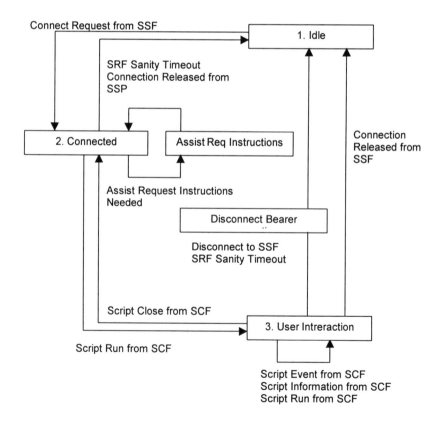

**Figure 2.7.2**
SRF state model

For backwards compatibility reasons, the SCF-SRF interaction based on the IN CS-1 information flows should still be supported. However, practically speaking, all interactions between the SCF and the SRF can be based on the execution of the scripts. The IN CS-2 compatible SRF State Model (SRSM) which supports the script execution is presented in Figure 2.7.2.

To further explain the idea of script execution on the B-IP, the User Interaction state can be decomposed in sub-states as it is depicted in Figure 2.7.3.

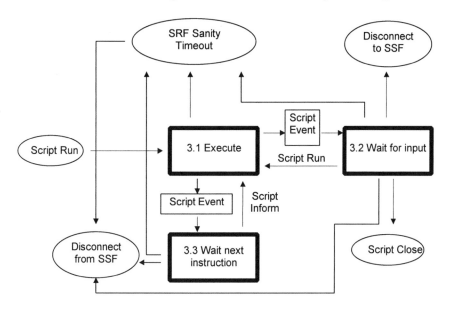

**Figure 2.7.3**
User Interaction sub-states

As it can be seen here, the initiation of a script execution is triggered by a 'Script Run' message which also triggers the allocation of the necessary SRF resources. A script execution action may be either terminated (normally or by an exception condition) or paused to acquire more information related to the script execution. In both cases a 'Script Event' is issued reporting the result of the script execution. In the case where additional information is needed for the completion of the script execution, this is provided by the means of a 'Script Information'. Finally, de-allocation of the SRF resources used during the script execution is performed following the reception of a 'Script Close' by the SRF.

## 2.7.4   Example Services

In the following paragraphs a real application of a B-IP used for supporting IN multimedia services will be presented based on a refinement of the architecture presented in the previous section. The multimedia services consist of an

Interactive Multimedia Retrieval (IMR) and a Broadband Video Conference
(B-VC) as they were introduced in Chapter 2.4 above.

## 2.7.4.1 B-IP Architecture

Following the architectural scheme presented in Figure 2.7.1, the B-IP
implementation which hosts the two multimedia services is presented in Figure
2.7.4.

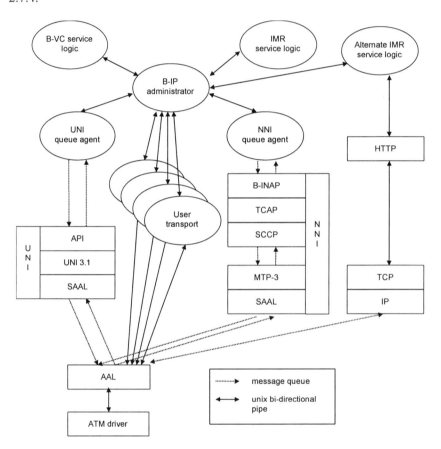

**Figure 2.7.4**
B-IP implementation architecture

The structural elements of this B-IP are the following:
* The *B-IP administrator (IPA)*. This module is the core element of the
  B-IP and communicates directly with the protocol stacks and the service
  logic instances. The communication with the SL instances is based on a
  standard interface allowing the easy introduction and integration of new
  SL modules on the B-IP.

- The *protocol stacks*. Figure 2.7.4, depicts three protocol stacks on the B-IP:
  - A B-INAP based protocol stack used for B-SCP/B-IP communication. This stack consists of the standard lower layers: SSCOP, NNI-SSCF and MTP-3 and the upper layers for IN: SCCP, TCAP and B-INAP. The B-INAP layer uses special extensions to the INAP to carry additional, service specific information.
  - A UNI based protocol stack used for interfacing the user terminals.
  - One or more transport protocol stacks used for performing message exchange between the B-IP and the user to send/collect information. Two stacks are implemented: a stack based on SSCOP and a stack for Classical Internet Protocol over ATM.
- The *service logic instances*. These contain the necessary functionality for the provision of IN services. The SL instances on the B-IP provide the means for performing information exchange between the user and the B-IP which is necessary for acquiring the information requested during IN service execution. A common part dedicated to the exchange of messages with the B-IP administrator is used in each SL which forms a unique communication way between the SL instances and the B-IP administrator.

## 2.7.4.2 B-IP Administrator

The B-IP Administrator (IPA) is the central module of the B-IP architecture. It has the following tasks:

- *Managing services and service logic instances*. The available services and their attributes are managed through a management interface. The IPA connects to the services, initializes them or shuts them down.
- *Handling communication with the SCP*. This communication is based on TCAP transactions which are executed through the B-INAP protocol stack. The IPA keeps track of all transactions and maps between the data format of the B-INAP stack and the service logic interface. This allows for an easy replacement of the B-INAP stack.
- *Handling communication with the UNI stack*. The IPA keeps track of all bearer connections. It maps between the UNI stack interface and the service logic interface, which allows for easy replacement of the UNI stack.
- *Initialization of user protocol stacks*. The IPA initializes the user protocol stacks depending on what stack is used: it either starts the SSCOP instances or manages the ATMARP table for TCP/IP connections. It keeps track of SSCOP instances and ATMARP table changes.
- *Handling user communication*. In the case of SSCOP based user communication, the IPA maps between the service logic interface and the user connection application protocol.

Another requirement is the logical decoupling of the different interfaces to allow easy replacement of modules. The structure of the B-IP, which is based on these requirements, is presented in Figure 2.7.5.

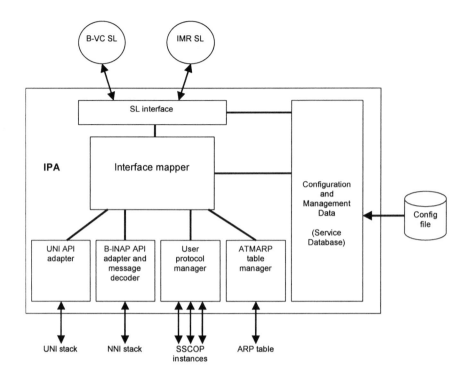

**Figure 2.7.5**
IPA structure

The structural elements of this IPA are the following:

- The *UNI adapter* module handles communication with the UNI protocol stack and maps between the UNI stack interface and the message based SL interface. Because the SRF never originates calls, this adapter originates two messages towards the Interface Mapper (IM): UNI_SETUP and UNI_RELEASE. It never receives any messages from the IM.

- The *B-INAP API adapter* handles communication with the B-INAP stack. It decodes and encodes message to/from the stack and converts them to the format used by the SL interface. This adapter also allocates various identifiers used by the B-INAP.

- The *User protocol manager* creates SSCOP instances and communicates with these instances, providing a mapping between the user protocol and the service logic interface messages. Each user connection is handled by another instance so the manager must keep track of these instances.

- The *ATMARP table manager* makes the necessary changes to the ATM driver's ARP table, when a new Internet Protocol connection must be set-up or an old one is to be deleted. It must keep track of the table entries, because there may exist more than one connection to the same Internet Protocol address.

- The *Service Logic (SL) interface* handles the channels toward the service logic instances by sending and receiving messages and by monitoring the start or disappearance of logic instances. The figure also shows two kinds of service logic for the two example services.
- The *service database* holds a list of available services together with their attributes. Each record of an active service (i.e. service with a running service instance) holds a list of running sessions, each of which in turn holds lists of TCAP transactions, user connections and ARP table references. The static part of this database is read from a disc file at startup and may be re-read at any time to change the set of available services.
- All messages go though the *Interface Mapper*. This part provides mappings between the various identifiers used by other modules. This mapping presents each service logic and, within each service logic each session, with a unique set of identifiers, preventing collisions in identifier use.

## 2.7.4.3 B-IP Administrator/Service Logic Interface

The IPA provides a standardized interface towards the service logic instances. Communication is based on bi-directional byte streams, so it is possible to use local (like UNIX sockets) or remote communication mechanisms (like TCP sockets). The latter may be necessary if the B-IP is implemented in a distributed way for maximum scalability.

The communication stream is used for the exchange of messages. Each message consists of a header, indicating the type and the length of the message, and a message type specific part, as well as an optional extension part. The extension part, if present, is not interpreted by the IPA but simply forwarded. It can be used, for example, to carry service specific data between the B-IP and the B-SCP or between the B-IP and the CPE.

There are four groups of messages, each group serves the communication between the service logic and another entity:

- *UNI messages.* These messages handle the communication between the UNI stack on the B-IP and the service logic: information flow is uni-directional, because the B-IP never initiates or releases connections. The messages inform the service logic, that a connection to the CPE was successfully established (including the transport layer) or released;
- *NNI messages.* These messages are for communication between the service logic and the B-SCP. They are carried through the B-INAP and correspond to the B-INAP information flows. All encoding/decoding (expect for the extension parts) is handled by the IPA;
- *User protocol messages.* They are dedicated to interaction with the user (in case of the TCP/IP transport protocol dedicated messages are not used). This interaction is based on synchronous request/response interaction with the possibility for asynchronous notification. Requests are always sent from the CPE. These requests arrive at the IPA/SL interface as USER_REQUESTs. The service logic handles the requests (possibly interacting with the B-SCP in the meantime to obtain further information) and responds with either an

USER_ERROR (if something went wrong) or a USER_OK, if anything was ok. The service logic can also notify the user at any time through USER_EVENTs. This is useful in multi-party services, when interaction with one user influences the CPE state of another user (if for example a user leaves a video-conference, it may be necessary to inform the other users);

- *Interface management.* The messages in this group have local meaning only. They are used to manage the interface between the IPA and the SL.

Figure 2.7.6 shows an example of communication betweena service logic and the IPA.

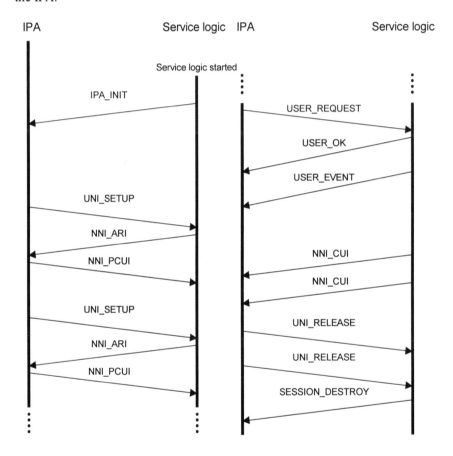

**Figure 2.7.6**
IPA/SL communication example

The example shows that when a new service logic is started, it registers itself with the IPA. Then a session with two user connections is created. Communication with the user takes place and the result of this communication is reported back to the B-SCP. At the end, the service logic receives the release messages for both connections and destroys the session.

# References

[1]  Anagnostakis C D, Patrikakis C Z, Prezerakos G N, Venieris I S, *An IN-based approach to the design of Interactive Multimedia Services over Broadband Networks : The VoD example*, Journal of Network and System Management, **5**, 329-350, 1997

[2]  ITU-T Recommendation Q.1228, *Interface Recommendation for Intelligent Network CS-2*, 1997

[3]  ITU-T Recommendation Q.1203, *Intelligent Network Global Functional Plane architecture*, 1995

# Chapter 2.8

# PRACTICAL EXPERIENCE WITH INTELLIGENT BROADBAND NETWORKS - THE INSIGNIA TRIALS

The project INSIGNIA has carried out practical trials of Broadband IN services in the context of an international ATM network involving three countries. This chapter describes the infrastructure which was used for these trials, illustrates the practical experiments which were carried out and summarizes some experiences from the trials.

## 2.8.1 Overview on the Project INSIGNIA

The project INSIGNIA (IN and B-ISDN Signaling Integration on ATM Platforms) started in September 1995 as the European research project AC068 within the ACTS (Advanced Communications Technologies & Services) Programme of the European Commission and is partially funded by this programme. The duration of the project is planned to be three years and it involves 14 partners out of eight European countries. This book was written by members of the INSIGNIA consortium. The INSIGNIA consortium includes research institutes as well as three network operators and five manufacturers of telecommunication equipment. The list of project partners is as follows:

- Siemens AG, Germany, co-ordinating partner;
- Italtel s.p.a., Italy;
- Telefónica de España, S.A., Spain;
- GPT Ltd, United Kingdom;
- Centro Studi e Laboratori Telecomunicazioni S.p.A (CSELT), Italy;
- GMD FOKUS, Germany;
- National Technical University of Athens (NTUA), Greece;
- Siemens Schweiz AG, Switzerland;
- Consorzio di Ricerca sulle Telecommunicazioni (CoRiTeL), Italy;
- Telecom Italia S.p.a., Italy;
- University of Twente, The Netherlands;
- Fondazione Ugo Bordoni, Italy;
- Siemens Atea, Belgium;
- Deutsche Telekom AG, Germany.

The objectives of the project are:

*Intelligent Broadband Networks*. Edited by I. S. Venieris, H. Hussmann
© 1998 John Wiley & Sons Ltd.

- to define, to implement and to demonstrate an advanced architecture integrating IN and B-ISDN signaling within the three years time frame of the project;
- to consider user requirements arising from new interactive application services;
- to enhance the existing infrastructure in several European countries (the so called National Hosts) to support trans-European switched ATM networking;
- to define, implement and demonstrate selected network services in field trials on National Host platforms with real user involvement;
- to achieve an integration of IN and B-ISDN concepts through contribution to relevant standardisation bodies.

The National Host structure has been set up by the Commission of the European Community (CEC) to provide co-ordinated access to advanced telecommunication networks for the ACTS projects. Each country has a National Host (NH), which is one organisation or a group of organisations (including network operators, universities and research departments) which has experience of Broadband networks and/or network infrastructure that can be made available to ACTS projects. The NHs also provide a national focus for ACTS activities.

The INSIGNIA project builds prototypes for the network elements and end systems of an Intelligent Broadband Network and develops example services for this network, following the technical approach which was laid out in detail in the preceding chapters. Using a fast prototyping approach, the project is structured into two phases. The first phase has delivered a set of prototypes for the first trial in summer 1997, the second phase leads to a second set of prototypes for the second trial in summer 1998 [1, 2, 3].

## 2.8.2   The Trials

### 2.8.2.1 The INSIGNIA Trial Network

The INSIGNIA trial network (Figure 2.8.1) is a trans-national network interconnecting three countries in Europe (Germany, Italy and Spain) by using the pan-European ATM facilities which are provided by the ACTS project JAMES, with five physical locations (Berlin, Munich, Milan, Turin and Madrid) [3].

The German island is composed by two different sites (Berlin and Munich), as is the Italian island (Milan and Turin). The Spanish island consists of just one site (Madrid).

The trials demonstrate three different example services (see Chapter 2.4 for detailed descriptions):

- A Video on Demand (VoD) service, which is extended in the second trial phase to a more comprehensive Interactive Multimedia (IMR) Service;
- A Broadband Video Conference (B-VC) service with high-quality audio and video;
- A generic Broadband Virtual Private Network (B-VPN) service which is demonstrated in connection with videotelephony applications and which is

extended in the second trial to a Broadband Virtual Intranet service (see Chapter 3.5).

These services are available in the various sites as follows: In the Berlin site, the B-VC service can be demonstrated and it can be considered as a remote user of the Munich site. In the Munich site, the network nodes and terminals for the demonstrations of both the B-VC and the B-VPN services are located. The German National Host is used for connecting both sites.

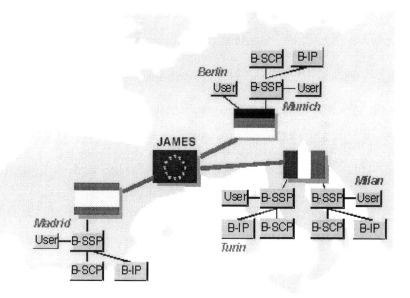

**Figure 2.8.1**
The INSIGNIA trial network

In the Italian island, the main focus is on VoD/IMR. In the Milan site, only the VoD service has been demonstrated. In the Turin site, both the VoD and B-VC services can be demonstrated. The Italian National Host (ITINERA) is used for connecting both Italian sites.

The Spanish island is composed by one site, Madrid, where the B-VC and B-VPN services were demonstrated during the first trial.

## 2.8.2.2 Service Trials and Demonstrations - First Trial Phase

Within the first trial, all three services mentioned above have been implemented and successfully demonstrated involving different islands.

First, during the ECMAST'97 Conference in Milan, Italy, the Video on Demand service has been demonstrated to a representative of the European Commission and to the public at the Italtel premises, connecting the Milan and Turin islands via the Italian National Host [4]. Figure 2.8.2 shows the configuration of the islands and the network elements.

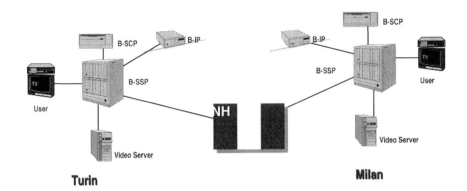

**Figure 2.8.2**
Video on Demand service trial configuration

Second, the B-VPN and the B-Video Conference services were demonstrated to a representative of the European Commission and thereafter several times at different occasions at the Siemens premises in Munich. Figures 2.8.3 and 2.8.4 show the configurations of these demonstrations.

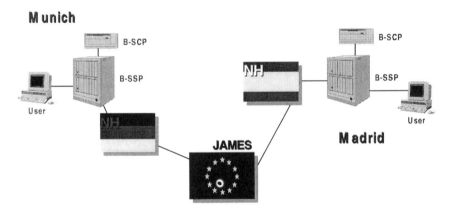

**Figure 2.8.3**
Broadband Virtual Private Network service trial configuration

For the Broadband Video Conference service several scenarios were prepared and successfully shown, like starting a conference with three local users (local in terms of the Munich island), releasing one user and including a (remote) user from the Berlin site, based on a connection via the German National Host, finally releasing another local user and establishing a trans-European connection via JAMES with a user in the Spanish island at Telefónica.

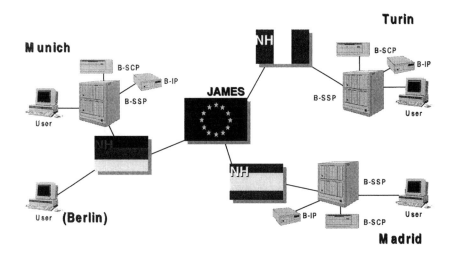

**Figure 2.8.4**
Broadband Video Conference service trial configuration

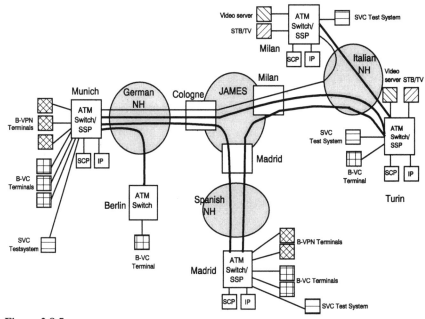

**Figure 2.8.5**
INSIGNIA network configuration (first trial phase)

During the visit of the European Commission various scenarios of the B-VPN service (between Madrid and Munich) could also be demonstrated like on-net to on-net calls, off-net to on-net and on-to off-net calls (or call attempts, respectively), illustrating audio and video performance and the capability of managing authentication features.

Figure 2.8.5 shows in more detail the configuration for the first trial. It illustrates not only the connection of the four national islands to the JAMES network via the National Hosts and the JAMES 'Points of Presence' (PoPs), but also the network elements and the CPE at the different sites: B-VPN Terminals (CIVA) in Munich and Madrid, Video Server and STB/TV sets in Milan and Turin for the Video on demand service, and B-VC Terminals in Munich, Madrid and Turin.

## 2.8.2.3 Service Trials and Demonstrations - Second Trial Phase

As a major improvement in the second phase of the project, users will be able to use network-wide IN services controlled by remote B-SCPs via a signaling network. This is a unique opportunity to validate on the one hand the concept of specialized, remote B-SCPs and on the other hand a signaling network for Broadband services.

The services used in the second trial are enhanced services of the first trial: IMR and B-VC use a unified B-IP, B-VPN combined with an IP over ATM feature growing to B-VI (Broadband Virtual Intranet). The B-VPN service moreover demonstrates the practical use of User-Service Interaction for features like a 'Follow-Me' function (call diversion).

The Milan island is going to extend its demonstrations in the second trial to include the B-VC service besides IMR. The German island will be enlarged by adding another remote site in Darmstadt provided by Deutsche Telekom.

## 2.8.3   Summary of Trials

When summarizing the project's experiences from the first trial one has to have in mind not only the objectives of the INSIGNIA project, as they have been planned several years ago during the proposal phase, but also the rapid changes in the area of high-speed networking, particularly on the background of the deregulation phase during the lifetime of the project.

The consortium is very grateful for the support by the European Commission to enable a project like JAMES. Without the JAMES ATM network infrastructure combined with the availability of the National Hosts it would have been more difficult to run the pan-European trials. Management of access to several involved national Public Network Operators facilities has been eased. Our experiences during many test sessions, ordered some weeks in advance by giving all requirements like time frame and bandwidth etc., gives the impression that co-operation in this area can be technically achieved, but leaves space for improvement. We hope the European Commission together with the Public Network Operators and upcoming new operators will continue to support the idea of a pan-European ATM network for research purposes.

INSIGNIA has been successful in showing that the approach of IN evolution towards Broadband multimedia systems is feasible. The INSIGNIA trial services demonstrate multimedia services, which are much advanced compared to existing Internet services (by using the Quality of Service guarantees of ATM). On the other hand, the INSIGNIA approach provides a smooth path of evolution for network operators, where most of the equipment and operating knowledge (for IN and B-ISDN) can be reused. The interoperability of different ATM switches (commercial and research types) in the context of a pan-European ATM network was one of the major issues of the first trial test and integration phase and concluded successfully.

In order to ensure a further proliferation of these ideas, INSIGNIA has already submitted some of its essential concepts to the ITU-T for standardization within IN Capability Set 3.

# References

[1] Koch B F, *Intelligent Networks,* Proceedings of IS&N'97 Conference, Como, Italy, May 1997

[2] von der Straten G, Totzke J, Zygan-Maus R, *Realization of B-IN Services in a Multinational ATM Network,* Proceedings of ICIN'96 Conference (International Conference on Intelligent Networks), Bordeaux, France, 1996

[3] Hussmann H, Koch B F, *IN and Broadband ISDN Signaling Integration on ATM Platforms - The INSIGNIA Project,* Proceedings of ComCon6 Conference 1997, Corfu, Greece, June 1997

[4] Hussmann H, Herrera F J, Pasquali R, Todorova P, Venieris I, Zizza F, *A Transnational IN/B-ISDN Integrated Network for the Provision of Multimedia Services,* Proceedings of ECMAST'97 Conference, Milan, Italy, May 1997

# Part 3

## NETWORK EVOLUTION AND INTELLIGENT NETWORK INTEROPERABILITY

# Chapter 3.1

# INTEROPERABILITY OF INTELLIGENT BROADBAND NETWORKS

In parallel to the Intelligent Broadband Networks other similar or radically different approaches are currently investigated for the rapid introduction of multimedia services to telecommunication networks. Most of them require significant changes in the system software, a requirement hardly accepted by the network operators. The main issue is not the extendibility of existing systems but rather the cost of a radical upgrade and the robustness of the final enhanced system. At this point the IN approach offers a readily available solution as it guarantees minimum modifications/enhancements to existing equipment Examples have been presented throughout this book: In Chapter 2.5 we show how a switching platform is enhanced to a Broadband Service Switching Point, and in Chapter 2.5, how a narrowband Service Control Point is transformed to a B-SCP.

Advanced signaling systems have started to appear from the early 90s (see Chapter 1.3). The main philosophy behind this is the same as that of any advanced programming language. The service software is split into a number of basic object classes. Then objects are associated according to a predefined set of rules and the service model is born. The idea is to establish a modular service architecture that theoretically allows the representation of any service independently of the underlying network. The final target could be customized services. This matching of signaling system design and service engineering is more vividly exhibited in the work of the Telecommunication Information Networking Architecture Consortium (TINA-C). TINA-C has allocated considerable effort for engineering a flexible and modular open architecture for distributed telecommunication applications. It is therefore interesting to investigate the relation between the Intelligent Broadband Network architecture and TINA and to show possible ways for the migration of IN towards TINA. Chapter 3.2 undertakes these tasks. To further show that the Intelligent Broadband concept is a future safe solution, in Chapter 3.3 the impact of the evolution of the Broadband signaling on IN is investigated. Technical answers are given to questions like what about IN if network signaling provides more advanced capabilities, is IN redundant in that case and if not what are the enhancements needed.

The compatibility of the Intelligent Broadband architectures with past and current time technologies is not less important, at least for network operators and service providers. Therefore one would expect a backwards compatibility analysis in which narrowband IN would be in a central position. Such a

*Intelligent Broadband Networks.* Edited by I. S. Venieris, H. Hussmann
© 1998 John Wiley & Sons Ltd.

discussion is presented in Chapter 3.4 where ways are shown for the interworking of narrowband SCPs offering typical narrowband IN services like Freephone and Virtual Private Network with B-SSPs.

Last but not least the case of Internet is discussed in Chapter 3.5. Internet has been widely accepted as the emerging technology for accomplishing the user needs in terms of data, audio and video communications. The Intelligent Broadband Network concept offers the functionality required for the secure operation of private networks that make use of the Internet concepts but are kept separated from the world-wide Internet for security reasons.

# Chapter 3.2

# COMPARISON AND INTEROPERABILITY OF INTELLIGENT BROADBAND NETWORKS WITH TINA

## 3.2.1 Introduction

The Telecommunications Information Networking Architecture Consortium (TINA-C) was founded in 1992 and currently comprises all major telecommunications equipment suppliers, network operators and computer vendors. Its main goal is the definition and validation of an overall consistent, flexible and open architecture for distributed telecommunications software applications, known as the Telecommunications Information Networking Architecture (TINA).

TINA is an integrated architecture for all possible kinds of services and networks including their control and management, but aims, in particular, at Broadband multiparty and multimedia as well as mobility and information services. TINA's vision is that all telecommunications, control and management applications will be very easy to create and maintain in a customized, evolvable and cost-effective way, independent of any underlying transport network, technology, protocol and geographical location. This will be achieved by several layering principles intended to separate the high-level applications from the physical transport network by means of a Distributed Processing Environment (DPE) which resolves heterogeneity and distribution. Furthermore, the stable control and management functions will be separated from the very dynamic and flexible service development and evolution of technology. TINA is a solution which takes into account the increasing globalization, liberalization and regulation activities, and therefore a fast-growing and changing marketplace with many players. This requires an architecture which satisfies all future stakeholder demands regarding an open telecommunications and information software market, that allows a large variety of co-operation and federation activities, and takes focus on security, reusability and reliability properties. TINA is based on a completely new approach by using the latest developments in software engineering, i.e. object-oriented modeling and Open Distributed Processing (ODP) techniques.

The realization of all these requirements would mean to overcome the current situation, where many different service creation, management and control systems, network protocols and switching platforms exist. All of them are

*Intelligent Broadband Networks*. Edited by I. S. Venieris, H. Hussmann
© 1998 John Wiley & Sons Ltd.

designed for specific purposes, run by different technologies, and use de facto and international standards.

Currently, TINA-C is entering a phase of consolidation of its results and specifications. When the work on TINA is finished, the communications industry will get a new, open software architecture, with a complete set of specifications, for evaluation. However, there is not much practical experience with this architecture, especially concerning its performance, stability and scalability in the time-critical network control area. This situation is expected to improve because, during the next few years, experiences and results from complex TINA prototypes will be made available by several research projects.

TINA turns out to be a very interesting and promising solution for the middle and long term. A key factor for the success of this architecture concept is the integration of existing systems and networks, like IN, B-ISDN, TMN and Internet/Intranet, into TINA. This means that TINA should be able to interact with those legacy networks, and use existing systems and standards as far as possible. The main reasons for this are the heavy investments made for them, and the wide-spreading of some of these technologies, e.g. for IN. Therefore, smooth migration and interworking scenarios have to be considered by the TINA world, and interactions with the standardization bodies, like ITU-T, and OMG, have to be consolidated as well.

Considering IN and B-ISDN, ITU-T enforces the specification of advanced capability sets towards multimedia and multiparty services. Especially in the IN area several research activities towards Broadband integration have been initiated (see Chapter 2.8). On the other hand, the TINA-C concepts attract growing attention in the communications business, especially the DPE which is expected to be a CORBA [1] Object Request Broker (ORB) implementation that has been standardized by OMG. This implies investigations of interoperability concepts between the Intelligent Broadband Network architecture, presented in this book, and TINA. This chapter includes migration and interworking scenarios between both domains and considers the IN as well as the B-ISDN areas.

Section 3.2.2 compares the TINA concepts with those in IN and B-ISDN. Migration and interworking strategies between TINA and Intelligent Broadband Networks are investigated at different levels of abstraction in Section 3.2.3. Finally, conclusions and an outlook on future prospects of TINA are given.

## 3.2.2    Comparison of TINA with Intelligent Networks and B-ISDN

The Intelligent Broadband Network covers IN as well as B-ISDN functions. Therefore, both domains will be considered below in order to provide the basis for interoperability.

### 3.2.2.1 Intelligent Networks and TINA

Both the IN and TINA architectures are designed for service control and management in bearer networks, but there are fundamental differences. Whereas the IN represents a function-oriented architecture (see Chapter 1.2), TINA is based on an object-oriented concept.

IN decomposes services into a set of reusable service features in terms of Service Independent Building Blocks (SIBs), and defines functional network elements for the distributed implementation of these SIBs. As shown in Chapter 1.2, IN separates switching and service control as well as service management, data and special resources by grouping these functions into different functional entities, namely the CCF/SSF, SCF, SDF, SMF and SRF. These entities interact by means of interaction diagrams which are physically realized via the Signaling System No.7 (SS7) network by a dedicated IN Application Protocol (INAP). Furthermore, the service control nodes remotely control the switches by use of a Basic Call State Model (BCSM).

TINA services are modeled, from the ODP computational viewpoint, by a set of Computational Objects (CO) which comprise logic, interfaces and operations. Within the TINA Service Architecture (SA) [2] a set of generic Service Components (SCs) has been defined, which can be used for service creation by means of specialization and aggregation. Services are developed from an evolving environment of components, and each product can build on the success of preceding products rather than starting from the same pre-defined component set, like SIBs, each time. The TINA-DPE supports the arbitrary distribution and migration of these COs, i.e. in TINA there is no need for specific network nodes with dedicated functions.

The TINA COs communicate via their operational interfaces by means of the DPE. These interactions correspond to the interaction diagram between IN functional entities. In contrast with IN (INAP over SS7), TINA does not emphasize or identify a role for SS7, i.e. it does not specify a particular Kernel Transport Network (KTN) which is used for the inter-DPE communication. But recently several contributions [3, 4] concerning the use of SS7 as Environment-Specific Inter-ORB Protocol (ESIOP) and the carriage of the General Inter-ORB Protocol (GIOP) via the SS7 network have been made (see Subsection 3.2.3.4).

Instead of the switch residing functions for call/connection control (CCF/SSF, BCSM), TINA adopts a management-oriented layered view of connectivity, known as Connection Management (CM) which has been defined within the TINA Network Resource Architecture (NRA) [5]. (Note that everywhere else in this book CM stands for Call Manager.)

Most existing IN service capabilities are related to flexible screening, routing and charging of calls. In TINA such functions are mostly related to the generic access session concept rather than to a specific service session, where the particular service logic for a TINA service resides. This means that there is no one-to-one mapping of IN services to TINA services, and of IN functional entities to TINA COs [6]. Finally, it should be emphasized that IN service control capabilities correspond to Computational Object functions of the Service Architecture within the TINA application layer.

Broadband services are more complex and require much more flexibility of service control than existing narrowband services. The current IN architecture lacks this flexibility, as it was initially designed to handle simple narrowband services, e.g. free-phone and premium-rate. Despite recent developments in the standardization bodies (ITU-T IN CS-2 and ETSI) towards Broadband, certain basic features, e.g. the BCSM and DP-mechanism, do not offer the required functions [7].

In Chapter 2.2 an innovative and flexible Intelligent Broadband Network architecture has been investigated. It is founded on existing systems and classical IN concepts (including BCSM and DPs), but maps those concepts into an abstract object-oriented model, which is the Switching State Model. Furthermore, this Switching State Model adopts an IN service session concept and gives the network view of an IN service session, many configurations of which are supported. Following this approach, the associated Broadband-INAP (B-INAP) protocol carries abstract operations on the Switching State Model (see Chapter 2.3). This object-oriented model can be compared with the TINA service session model, the Service Session Graph (SSG), because both represent an abstract view of a (TINA or IN, respectively) service session, and are modeled by the Object Modeling Technique (OMT) from the ODP information viewpoint. Consequently, interworking between the IN architecture presented in Chapter 2.2, and TINA, is expected to be easier than between classical IN and TINA.

## 3.2.2.2 B-ISDN and TINA

In current B-ISDN, the signaling refers to the real-time control-plane capability to supervise a call and its connections in terms of set-up, modification and release of connections in the user plane. B-ISDN connection management is performed on a link-by-link basis by peer-to-peer-based signaling protocol procedures (see Chapter 1.3).

In TINA the Connection Management (CM) is very much different from the B-ISDN solution. The concept of signaling corresponds to interactions between Computational Objects of the Service Architecture, COs of the Network Resource Architecture, and between COs of both architectures, by means of the DPE and KTN. The purpose of these interactions is the request for connectivity, which is broken down in a hierarchical top-down approach by the TINA-CM. Consequently, the CM provides a management-oriented layered view of connectivity, similar to the TMN concepts. At the highest level of abstraction, the service session in the Service Architecture, connectivity is expressed by means of stream binding between stream interfaces of certain COs. This kind of relationship is represented by the TINA Service Session Graph. In the TINA-Network Resource Architecture, connectivity is indicated by the Logical Connection Graph (LCG) within the communication session, the Physical Connection Graph (PCG) within the connectivity session, trails in a layer network, subnetwork connections in a subnetwork, and, at the lowest level, by actions performed by TINA-conform network element software, for instance, in order to assign a switch port [5]. This approach assumes an open TINA Object Definition Language (ODL) interface to the Connection Management for all network elements.

In existing B-ISDN systems the call and connection control domain is designed in a monolithic way, but as shown in Chapter 1.3 efforts are made in the standardization bodies to separate them in order to allow a call association without connections, e.g. for negotiation and look-ahead signaling capabilities. The TINA-CM is already based on that principle. The concept of call has been translated into the service session, and the connection concept is supported by the overall TINA-CM connectivity method. This separation has more the nature of a differentiation between functions in the service layer and the network resource layer of TINA. This finds its computational representation in the separation between service and communication session, and therefore between the TINA Service Session Manager (SSM) and Communication Session Manager (CSM) Service Components. From the information viewpoint this concerns the separation of Service Session Graph and Logical Connection Graph.

In B-ISDN, signaling is separated into User-Network-Interface (UNI) and Network-Node-Interface (NNI) signaling. Both are reflected in the SA and NRA by support of the intra-domain and inter-domain reference points defined by TINA-C in relation to the TINA business model [8]. The classical user-to-user signaling and the bearer-independent signaling concept is supported by the TINA architecture as well [9].

When considering a pure B-ISDN service development, this draws another major difference between B-ISDN and TINA architecture: Currently the knowledge of the underlying technology, operating system, protocol, etc., is necessary. This is not valid for a TINA-conform service creation method.

Another aspect of a B-ISDN/TINA consideration is the potential use of B-ISDN as Kernel Transport Network for inter-DPE communication. Since the KTN is used to enable the distributed inter-object communication between different DPE nodes via operational interfaces, these interactions could be implemented by the remote operation facilities of B-ISDN, i.e. the TCAP in SS7 and the GFP (Generic Functional Protocol) at the UNI. Furthermore, it seems to be useful to employ the B-ISDN, i.e. a non-TINA-CM function, to perform stream binding between TINA Computational Objects for the provision of connectivity. This replacement of the TINA Connection Management by B-ISDN signaling functions is conceivable at different layers of the TINA network model [5] (see Subsection 3.2.3.3).

## 3.2.3   Interoperability of TINA with Intelligent Broadband Networks

### 3.2.3.1 General Issues

To investigate interoperability between TINA and Intelligent Broadband Networks, two major issues have to be taken into account: migration and interworking. Whereas migration is related to a smooth introduction of TINA concepts in current telecommunications networks, interworking is related to the cooperation between TINA-conform and conventional systems by means of particular adaptation or interworking units. Migration has to be considered because there are a lot of existing legacy systems in the telecommunications industry for which heavy investments have been made by the telecommunications network operators and system companies. Consequently, the success of TINA depends largely on well-investigated migration scenarios which are based on slowly expanding islands of TINA-conformity. The result of migration is a full TINA-compliant network. As long as this goal is not met, interworking between legacy and TINA systems will be necessary. Since TINA is a solution for the middle and long term, interworking is expected to be required for a long time [6].

Interworking between TINA and non-TINA systems is possible and has to be considered at different horizontal levels of abstraction, namely the TINA application and DPE/KTN levels. At the application level, the vertical separation of service and network resource layers should be taken into account.

At the Service Architecture level interworking between TINA Service Components and IN Functional Entities will be considered for several stages of migration. This involves the specialization of the TINA Service Components according to the required IN-like function, and the interworking between IN B-INAP operations and TINA-ODL operations at the operational interfaces of appropriate COs. With respect to Chapter 2.2, the model conversion between the IN Switching State Model and the TINA Service Session Graph is concerned here, too.

At the Network Resource Architecture level the use of B-ISDN signaling capabilities to replace or support the TINA Connection Management will be investigated. This can be done for several migration steps and includes the mapping between B-ISDN signaling messages and operations on TINA-CM objects.

At the DPE/KTN level the possibility to use existing B-ISDN signaling capabilities in order to support the TINA Kernel Transport Network will be discussed. This includes interworking between B-ISDN SS7 protocols (e.g. TCAP) and TINA inter-DPE protocols (e.g. CORBA-based GIOP) at the NNI as well as between Generic Functional Protocol and those inter-DPE protocols at the UNI. Again, several migration steps, from initial interworking to the use of B-ISDN as KTN, can be considered. Even the provision of a B-ISDN based DPE, that uses B-ISDN signaling to provide operational as well as stream binding, is worth to be analyzed.

### 3.2.3.2 Migration

To investigate migration scenarios from Intelligent Broadband Networks towards TINA, the advantages of an introduction of TINA solutions into IN have to be considered. They are mainly related to:

- The possible rationalization of service aspects, like the integration of service management and control;
- The ability to extend service-related capabilities, e.g. several points of control;
- A higher level of interworking between applications running on top ofDPEs;
- The scalability of the service platform and the vendor independence;
- The evolvability and reusability of the service software;
- The cost-effective and fast-to-market service creation.

Therefore, the TINA concepts could be used in the following main IN areas [10]:

- *Service Management* - because of the lack in standardized IN management solutions and the integration of management and control;
- *Service Data* - on account of the advantages of object-orientation;
- *Service Control* - due to flexible access and service session capabilities.

Since a migration from IN to TINA will be based on a step-by-step replacement of IN Functional Entities (FEs), these entities will be modeled in TINA as a set of interacting Computational Objects. This process is referred to as TINArization of IN FEs [6]. From the technical point of view, this is associated with an extension of the TINA-DPE (ORB implementation) across a growing number of network nodes, until all switches, special resources, databases and CPEs will be involved. This results in the replacement of the existing static interfaces and protocols for service control and management (e.g. INAP and CMIP) by flexible DPE/CORBA mechanisms (e.g. GIOP, ESIOP).

The whole migration process requires the definition of Adaptation Units (AUs) which support the communication between the existing IN and the new TINA part. These Adaptation Units are, in fact, interworking units that provide the necessary protocol conversion, e.g. between B-INAP operations and TINA CO interface operations, and model adaptation, e.g. between the TINA Service Session Graph and a certain IN Switching State Model (see Chapter 2.2).

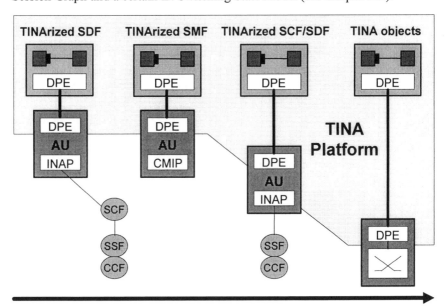

**Figure 3.2.1**
Migration Steps from IN towards TINA

Figure 3.2.1 shows possible evolution steps from IN towards TINA [10]. The stand-alone TINArization of the IN-SMF and SDF has to be taken into account, but it is quite unrealistic [6]. This is because these migration steps decouple service logic, service management and service data, which is not in line with the TINA approach concerning object-orientation, and control and management integration.

Therefore, a more realistic scenario is the integrated TINArization of SMF, SCF and SDF [10]. Only the SSF/CCF will remain IN-compliant because this IN Functional Entity has already been deployed in a lot of switches on a global basis. This scenario allows the network operators to keep the existing BCSM and INAP interfaces as long as required, whereas service control and management are implemented as TINA Computational Objects on top of the DPE. This may be realized in a TINA-based IN node, or by a set of distributed TINA nodes, according to the concepts of distribution hiding from the applications. Adaptation Units are required to perform interworking between the TINA COs implementing IN-SCF/SDF and SMF functions, and the IN-SSF. This will result in interworking between INAP and CMIP operations, and TINA Computational Object interface operations, on the other hand.

The last step would be the replacement of the IN-SSF/CCF. This assumes a programmable Broadband switch which offers a TINA-conform CO interface to control the switching functions. In the TINA sense, TINA-compliant network element software runs on that switch. This driver software is controlled by service session objects of the SA, and Connection Management objects of the NRA, via the DPE. Then, the INAP interface would no longer be necessary, because the switches can directly perform the operations invoked by the TINA Computational Objects. With respect to Intelligent Broadband Networks, this step would also mean replacing the B-ISDN signaling functions of the CCF by the TINA-CM. Since the B-ISDN signaling can be seen as an accepted and real-time proved standard (ITU-T and ATM-Forum), which is available as a product on the market that is implemented in many existing ATM networks, the sudden introduction of the whole Connection Management would be a costly solution. Moreover, further investigations of the CM's real-time performance will be necessary. So the aim of a migration from B-ISDN to TINA should be to use the existing signaling standards as long as possible, instead of replacing the B-ISDN signaling completely from the beginning [6]. According to the above considerations, a smooth substitution of the classical signaling functions by TINA-CM layers seems to be a promising way. This implies the conversion between B-ISDN signaling messages and the interface operations of the appropriate Connection Management objects at the communication session, connectivity session, layer network, or subnetwork level, depending on the migration stage.

The TINA-SRF has not been considered in the previous scenarios because its purpose is more related to resource than to service control. If needed, it can be modeled as a Service Support Component (SSC) in the TINA Service Architecture [2]. This component can own operational as well as stream interfaces which can be bound by using the TINA-CM. A second way is to support the SRF by another TINA business role, like the third-party service provider [8].

### 3.2.3.3 Application Level Interworking

The previous migration scenarios show that the most interesting interworking issue concerns the communication between SSF and SCF. In the following, it is assumed that the IN service logic and data will be realized by appropriate TINA COs. Therefore, Interworking (IW) between the IN-SSF (in the IN domain) and a TINArized SCF (in the TINA environment) will be needed. The Adaptation Unit responsible for this has to provide a mapping between B-INAP operations and TINA CO interface operations. This AU looks from the IN side like a SCF and from the TINA side like a TINA end-user system comprising several Computational Objects. Thereby, the Adaptation Unit enables the establishment of an association between IN-SSF and the TINA Service Session Manager (SSM) Computational Object, where the service logic resides. So the TINA-SSM controls the IN-SSF and the related CCF via the Adaptation Unit. The function of this AU depends on how IN service capabilities are modeled within the TINA environment. Two options can be taken into consideration [10]:

1. The calling and the called parties do not use a TINA end-system and are not modeled as TINA users. This is likely to happen at the beginning of migration.
2. One or more parties use a TINA end-system and are modeled as TINA users. This will occur at a later migration stage.

Below, both configurations will be investigated in more detail.

### Case 1

In the first case, a call and its connections will entirely be established by the IN-SSF/CCF under control of the TINA-SSM, supported by the Adaptation Unit. Therefore, no communication session is needed. This interworking configuration is shown in Figure 3.2.2.

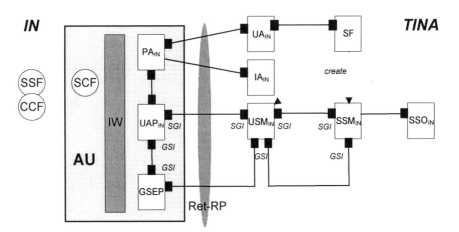

**Figure 3.2.2**
IN-SSF/TINA-SCF Interworking Configuration 1

The IN-like service capabilities can be completely modeled as TINA service session COs, derived from the TINA Service Components. This includes a User Application (UAP$_{IN}$) associated with a User Session Manager (USM$_{IN}$), and a Service Session Manager (SSM$_{IN}$) which makes use of a service-specific Service Support Object (SSO$_{IN}$), e.g. a database [6]. The TINA UAP Service Component has been mapped on two Computational Objects, the UAP$_{IN}$ and the Generic Session End-Point (GSEP). The UAP$_{IN}$ covers all service-specific aspects, whereas the GSEP encapsulates all generic service session aspects, e.g. the negotiation of the service session control model and interface [2].

Since no parties are modeled as TINA users, no user-specific access session is necessary. So generic access session related Computational Objects are the Provider Agent (PA$_{IN}$), the Initial Agent (IA$_{IN}$) and the User Agent (UA$_{IN}$), realized as anonymous UA [2]. When considering this configuration, the AU probably relates to the Retailer (Ret) reference point of the TINA business model [8]. The Generic Service Interface (GSI) at the service session oriented COs is responsible for the service session model and interface negotiation, whereas the Service Session Graph Interface (SGI) enables operations on the SSG, in case the TINA-SSG has been selected as service session model.

Since no communication session is necessary, the SSM$_{IN}$ does not create a Communication Session Manager, and therefore does not need to perform a mapping between Service Session Graph and Logical Connection Graph [5].

## Case 2

The difference from case 1 is that TINA-conform end-user systems are available and can be modeled as TINA users, as shown in Figure 3.2.3.

So user-specific access sessions which use a named UA will be necessary at the TINA user side B. Another important issue is that the TINA Connection Management has to be used to establish connections, i.e. to perform stream binding in the TINA domain. For this reason, the SSM has to create a CSM, i.e. the provider service session has to invoke a communication session, and a Service Session/Logical Connection Graph conversion will be needed. The CSM will invoke a connectivity session including the Connection Coordinator (CC) related objects, and thus create the Physical Connection Graph. (Note that everywhere else in this book CC stands for Call Control with the exception of Subsection 1.2.4.2 where it stands for Call Configuration in the context of CS-2 IN-SSM). Finally, the Layer Network Coordinator (LNC) related COs instruct the Connection Performers (CPs) to establish the physical connection between the Network Trail Termination Points (NWTTP) of the TINA Layer Network (LN), only one of which is assumed here. This means that all principles of the TINA Network Resource Architecture [5] apply. Since the calling user A is located in the IN domain, which is provided here because of the TINArized SCF scenario, whereas the called user is a TINA user, some kind of bearer connection bridge will be necessary in the Adaptation Unit [10]. This is because one part of the connection (user A - AU) will be established by the IN-CCF/SSF, and the other part (AU - user B) will be established by the TINA-CM (see Figure 3.2.3).

Therefore, the AU has to provide additional Computational Objects to support the Connection Management, i.e. the Terminal Communication Session Manager (TCSM) and the Terminal Layer Adapter (TLA). Consequently, the AU also relates to the Terminal Connection (TCon) reference point of the TINA business model [8] between retailer and connectivity provider.

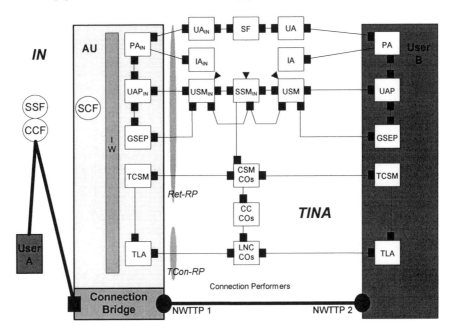

**Figure 3.2.3**
IN-SSF/TINA-SCF Interworking Configuration 2

In the last interworking configuration it is assumed that the full TINA Connection Management has been implemented, but regarding a smooth migration from the underlying B-ISDN towards TINA (see Subsection 3.2.3.2) this is likely to happen step by step, as considered below.

A first scenario could allow the TINA service objects to request connections in the network between two or more parties by means of B-ISDN signaling. Here, the TINA service session objects can directly invoke the bearer connection establishment by means of a special interface at a number of switches in the connectivity provider domain. To establish the needed connections, a network initiated call/connection set-up will be necessary. This corresponds to the third party call control B-ISDN signaling capability, which has not yet been standardized, neither in ITU-T nor in ATM-Forum, but is under study.

There are several ways for a switch with a TINA service object access interface to establish connections between the different parties involved in a service session. One solution is given if the TINA service objects send their connection request to the Local Exchange (LEX) of the initiating user. It requires that all LEXs include the TINA service object access interface, and therefore also a DPE kernel. In this scenario an Access Unit is needed at the LEXs to perform a conversion between the LCG operations of the TINA-SSM service object and the appropriate signaling messages for the third-party call set-up [7]. Within this configuration, connections can be directly established, modified, maintained and released by the TINA service session. Since this connection management function resides in the connectivity provider domain, there is a high level of security. Major drawbacks are the additional signaling requirements as well as the heavy changes at the LEXs.

Another way to implement this scenario is to place the service object access interface, the DPE and the Adaptation Unit in a few Transit Exchanges (TEXs). Then, a B-ISDN signaling mechanism is needed to forward the set-up request to the appropriate LEX. Besides the third-party call control signaling, the proxy signaling capability standardized in ATM-Forum UNI 4.0 [11] could be used [6]. In the latter case, a proxy signaling agent in the AU takes over the signaling work on behalf of the user. Following the TEX-related way would avoid the need to install the Adaptation Units at all LEXs.

The next B-ISDN interworking scenarios are very similar to the previous one. The main difference lies in the fact that a part of the TINA-CM will be introduced in the TINA nodes. This influences the design of the Adaptation Unit in the switches. A first step would be the introduction of the TINA communication session objects. Here, the AU has to convert between Physical Connection Graph operations and B-ISDN signaling messages. Then, step by step, the TINA connectivity session, the layer network and the subnetwork objects can be included [5]. Such a strategy allows the smooth introduction of the TINA Connection Management as well as extensive performance studies of the Network Resource Architecture, but, on the other hand, it implies a frequent re-design of the Adaptation Unit and the CM object access interfaces in the switches.

## 3.2.3.4 DPE and KTN Level Interworking

In the previous chapter, interworking between INAP operations and TINA Computational Object interface operations at the TINA application level has been considered. At a lower level this requires the interworking between the SS7 network, which provides the basis for the INAP, and the Kernel Transport Network, which provides the means for the inter-DPE communication. Since OMG's Object Request Broker is expected to be the implementation basis of the TINA-DPE, the inter-DPE communication will be realized by means of inter-ORB protocols, like GIOP and its Internet Inter-ORB Protocol (IIOP) specialization [1]. This results in the investigation of the interworking between inter-ORB protocols and SS7 protocol suite. These issues are currently under evaluation in OMG [3,4].

To ensure the maintenance of existing signaling protocols for the communication between telecommunication systems, two evolution steps have to be considered carefully (see Figure 3.2.4):

1.  Bridging/Interworking between SS7 and ORB protocols.
2.  Implementation of the GIOP on top of SS7 and the use of SS7 as ESIOP.

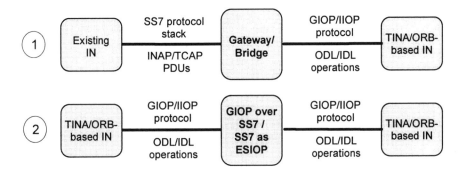

**Figure 3.2.4**
CORBA/SS7 Evolution Steps

In the first case, existing IN equipment, e.g. a SSF, has to communicate with TINA/CORBA IN-like implementations, e.g. a TINArized SCF. The TINA/CORBA objects/interfaces are defined by Object Definition Language/Interface Definition Language (ODL/IDL) specifications. Currently CORBA uses IDL, but the integration of TINA-ODL and CORBA-IDL is going on in OMG and ITU-T. The existing IN implementations use standardized INAP remote operations embedded in TCAP PDUs, which are transported across the SCCP/MTP stack in SS7. Signaling applications using TCAP, such as INAP, are defined as Remote Operation Systems Element (ROSE) user - Application Service Elements (ASE). These ROSE ASE specifications are written in Abstract Syntax Notation No.1 (ASN.1), and use the various constructs and concepts of ROSE [12] which is basically founded on a request/reply mechanism for remote operation invocations. ODL/IDL is based on the same paradigm, but uses another type of description in a programming language-independent manner. Consequently, the bridge/gateway (AU) has to perform an interworking between ROSE ASE constructs written in ASN.1 (e.g. INAP) and the corresponding TINA/CORBA constructs in ODL/IDL.

The second case concerns the communication between IN implementations which has already been TINArized. This means that no external bridge/gateway is necessary, and the SS7 is used as Kernel Transport Network, because it is a wide-spread and highly reliable signaling network for the time-critical communication between network elements. In this scenario the GIOP protocol messages have to be mapped on TCAP PDUs. Therefore, the TCAP/SCCP/MTP protocol part of the SS7 can be used here as ESIOP.

From the B-ISDN point of view, a signaling-based TINA-DPE seems to be an interesting issue. This implies the use of existing Broadband signaling capabilities to perform the operational as well as the stream binding of TINA objects, i.e. the use of B-ISDN as KTN and Transport Network (TN) for TINA.

The explicit binding of stream interfaces in TINA can be based on a DPE stream channel concept, where regular B-ISDN signaling messages are used in order to set-up, modify and release ATM bearer connections for multimedia information exchange [13].

The principle of operational binding in TINA is strongly related to the operation call principles defined by OMG. For the inter-DPE, i.e. inter-ORB, communication currently only the TCP/IP based realization IIOP of the GIOP has been standardized. The IIOP could serve as the basis for the intra-domain (e.g. inside a local network) operational interactions. If we consider the use of B-ISDN as Kernel Transport Network for the inter-domain object interactions, the IIOP could be used by means of an Internet Protocol over ATM access to the B-ISDN network. But here only the pure ATM TN will be used. A more appropriate solution would be the specification of new GIOP specializations that are adapted to current B-ISDN signaling standards at UNI and NNI (see Figure 3.2.5).

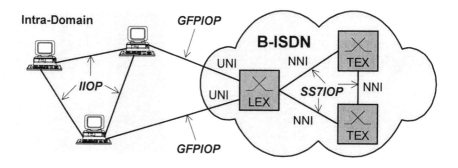

**Figure 3.2.5**
Inter-ORB Protocols using B-ISDN as KTN

At NNI the CORBA operation calls have to be mapped on TCAP protocol messages inside the SS7 signaling stack (SS7IOP), as previously mentioned with respect to current OMG activities. At UNI the use of the Generic Functional Protocol (GFP), according to ITU-T Recommendation Q.2932.1 [14], seems to be useful for this task. This is because it defines common mechanisms for the exchange of information between B-ISDN applications, and since it is specified as a realization of ROSE which supports interactive applications in a distributed open systems environment. The GFP specialization of the GIOP is referred to as GFPIOP, and is currently under evaluation in several projects [15].

## 3.2.4 Further Perspectives

The main advantage of TINA is related to the service creation, control and management areas, for which TINA provides an open integrated solution.

Services are expected to be built from an evolvable library of reusable Service Components in a fast and cost-effective way, with no need to have knowledge about the distribution and heterogeneity of the underlying network platforms. This is resolved by the DPE which is located between the service application layer and the transport network. So TINA is able to offer the advantage of a programmable distributed intelligent service layer, whose software is widely compatible and would be used by a lot of players in the future telecommunications market, and which can be exploited across different hardware platforms. Hence, the TINA service concepts are well suited to the upcoming globalization and liberalization in the field of communications, and offer new business opportunities (e.g. for service or software components).

Furthermore, TINA can improve management applications in general, especially the service management. Whenever no tight real-time constraints exist, the open hierarchical approach of the TINA architecture, which is similar to the TMN approach, is likely to provide a solution to uncovered or inflexible areas in management, e.g. for a unified service management architecture. The close relation to the TMN principles allows an easy co-operation between both management architectures [6]. All advantages mentioned above have to be seen from the viewpoint of today's telecommunications systems.

The major drawback of TINA is that it has concentrated on defining a new architecture and not sufficiently addressed the legacy issues of existing telecommunications systems. Moreover, it proposes a completely new network and connection management architecture, which is recognized as being slower than existing solutions [6]. This makes this architecture difficult to be adopted by the Telecommunications as well as the Internet world. Since the TINA specifications have not yet been finished, extensive practical performance studies of the whole TINA architecture will only be available during the next few years from complex prototypes. The TINA specifications have now reached a high level of detail in the definition of general principles and solutions, especially for the software design. The key step for the success of TINA is the detailed specification of components that follow the TINA approach and can be used at the same time in real networks supporting existing or near-term network interfaces, e.g. session manager objects with INAP-related interfaces, or Connection Management objects with B-ISDN signaling interfaces. This means that TINA-compliant components have to be easily adaptable to legacy systems. Such a way is necessary in order to save the heavy investments in existing networks for the long term. On the other hand, this may require the standardization of new signaling features in existing systems in order to support TINA, e.g. the third-party call set-up signaling capability. In either case, the survival of TINA will require the definition of very smooth migration and interworking scenarios from existing telecommunications systems, like IN and B-ISDN, towards TINA.

Although TINA was originally designed for Broadband multimedia and information services, the use for the information services seems to be very unrealistic [6]. This is mainly because those services appear to be provided more easily by the use of the wide-spread and fast-growing Internet. In this domain, the introduction of TINA seems to be very difficult since it is a quickly expanding market. Moreover, Internet is quite distant from the TINA architecture, because the Internet service management is located at the terminals or is supported by the Internet Protocol. On the one hand, Internet is a threat to TINA: Broadband networks are not spreading as fast as was forecast, and interactive multimedia services are provided by the Internet in an easy and cheap way, regardless of the large delays and the low QoS. This fact will become less important considering the new Internet Protocol version 6 and the RSVP which will guarantee a certain Quality of Service. On the other hand, TINA could learn from the Internet experience because many Internet technologies could be used to support TINA concepts, like the use of Java for building distributed applications, or the concept of the network computer.

All previous issues have shown that TINA is an important topic in today's telecommunications world. A large number of international and national research projects have been initiated in order to evaluate TINA, especially its performance and possible introduction into present telecommunications networks. Examples are the ACTS project RETINA as well as the EURESCOM projects P508 and P715. Furthermore, standardization bodies currently deal with TINA-related issues, e.g. the OMG concerning CORBA/IN integration. Moreover, OMG's CORBA is becoming a widely accepted technology. Because of this increasing importance of TINA, the Intelligent Broadband Network has to define its position with respect to TINA in terms of investigations regarding the interoperability between both architectures. This has been aimed at in this chapter. It has shown that a co-operation between the TINA and Intelligent Broadband Network concepts is possible, in principle, by means of Adaptation Units. Migration steps and interworking scenarios at different levels of abstraction have been given for the IN and B-ISDN areas. The innovative IN Switching State Model defined in Chapter 2.2 is likely to ease interworking between both domains, in contrast with conventional IN where very complex Adaptation Units are needed [6].

## References

[1] OMG, *Common Object Request Broker Architecture and Specification*, Updated Revision 2, July 1996

[2] TINA-C, *Service Architecture*, Version 5.0, June 1997

[3] OMG, *Intelligent Networking with CORBA*, Draft 2.0, White Paper, October 1996

[4] OMG, *RFI on issues concerning intelligent networking with CORBA*, telecom/96-12-02, DTC Document, November 1996

[5] TINA-C, *Network Resource Architecture*, Version 3.0, NRA_v3.0_97_02_10, February 1997

[6] EURESCOM, *Migration Strategy and Interworking with Legacy Systems,* Project P508 Deliverable 2, http://www.eurescom.de/public/deliverables/dfp.htm, April 1997

[7] EURESCOM, *Initial Assessment of the Options for Evolving to TINA*, Project P508 Deliverable 1, http://www.eurescom.de/public/deliverables/dfp.htm, December 1995

[8] TINA-C, Business Model and Reference Points, Version 4.0, May 1997

[9] TINA-C, *Connection Management Architecture*, TB_JJB.005_1.5_94, March 1995

[10] Herzog U, Magedanz T, *From IN toward TINA-Potential Migration Steps*, Proceedings of IS&N'97, Como, May 1997

[11] ATM-Forum, *UNI Signaling 4.0*, Approved Specification, July 1996

[12] ITU-T Recommendation X.880, *Data networks and open systems communication. OSI applications. Remote operations*, 1994

[13] Fischer J, Holz E, Kath O, Geipl M, Vogel V, *Provision of TINA Object Interaction through B-ISDN in ATM Networks*, Proceedings of GLOBECOM'97, Phoenix /Arizona, November 1997

[14] ITU-T Recommendation Q.2932.1, *Broadband Integrated Services Digital Network (B-ISDN). Digital Subscriber Signaling System No.2 (DSS2). Generic Functional Protocol Core Functions*, 1996

[15] Holz E, Kath O, Geipl M, Vogel V, Lin G, *Introduction of TINA into Telecommunication Legacy Systems*, Proceedings of TINA'97, Santiago, November 1997

# Chapter 3.3

# RELATIONSHIP OF INTELLIGENT BROADBAND NETWORKS WITH PURE BROADBAND ISDN

Intelligent Networks is not the only approach to support the connectivity for multimedia services. Another approach is the enhancement of signaling capabilities. For example, B-ISDN is evolving towards a network providing complex call configuration without IN involvement. The Intelligent Broadband Network concept is not controversial to these future enhancements but instead it can be seen as an intermediate step of harmonizing an advanced B-ISDN with IN, as will be shown in the following pages.

The architecture for the integration of IN and B-ISDN as described in the previous chapters has been mainly tailored to the B-ISDN Signaling Capability Set 1 (CS-1) with some features coming from CS-2. In this chapter the Intelligent Broadband Network approach will be first compared with the 'target' B-ISDN known as CS-3, showing that the integration with IN overcomes some drawbacks of the CS-3. Then it will be shown how the IN-based Broadband architecture is future-proof with respect to the enhancements of B-ISDN capability which are targeted to the short and medium period. Even if it is not clear when (and if) the most advanced control features will be available at the B-ISDN level, the Intelligent Broadband Network can profitably cope with these features.

The requirements for long term evolution of the Broadband network has been settled some years ago. The target set of network capabilities to be included in the B-ISDN (as defined for example in the CS-3 requirements) is very ambitious: very complex call configurations are handled by the B-ISDN control, full separation of call and connection control is exploited, complex special resources are controlled by the network, and so on. In this long term vision all the service functionality resides within the B-ISDN that has the complete control of any interaction among terminals involved in the call and possibly provides the required specialized resources. Sophisticated signaling mechanisms are needed to support this approach [1] which allows the network to have a full vision of the call, to guarantee the needed grade of service, and to make an efficient usage of precious network resources. On the other hand, there are some major drawbacks. The control procedures and signaling protocols become extremely complex. To introduce new services or to modify the existing ones requires a lot of standardization, implementation and deployment effort (i.e. software in all nodes has to be modified). The Intelligent Broadband Network architecture tries to reach a high degree of network control with minimum complexity and higher flexibility, by fully exploiting the advantages of the IN architecture. Many of the requirements which led to define the target B-ISDN CS-3 can be fulfilled by

*Intelligent Broadband Networks*. Edited by I. S. Venieris, H. Hussmann
© 1998 John Wiley & Sons Ltd.

integrating the IN service control on top of a simpler underlying Broadband network. In other words an Intelligent Broadband network architecture as considered in this book can be as powerful as B-ISDN CS-3, but it should be much more flexible and simpler to realize.

Coming to a shorter term perspective for the B-ISDN evolution, two major topics will be considered: the point-to-multi-point calls/connections which are already specified in the CS-2.1 recommendations and the multi-connection calls, which belong to future CS-2.2. It is worth spending some words on the impact of the evolution of B-ISDN capability onto the Intelligent Broadband architecture. The architecture is really open to future evolution. Two different mechanisms are available to support this evolution. In some cases, the enhancements of B-ISDN could be hidden to IN service logic and handled at the SSF level, where the mapping between the session domain and the connection domain is performed (see Chapter 2.2). This allows the 'stability' of the INAP SSF-SCF interface and it is an advantage of the session concept that 'uncouples' the service control level from the control of the underlying network. For example the introduction of multi-connection calls can be accomplished in this way, as it will be shown below. In other cases, the session model can be easily enhanced with the addition of new objects and/or new attributes: the choice of the object orientated session model pays back in terms of modularity. The new capabilities can therefore be directly offered to the IN service logic control. This last approach will be followed for the introduction of the B-ISDN point-to-multi-point call architecture, as it will be described hereafter.

## 3.3.1   Architectural Enhancements for Point-to-Multipoint Calls

A point-to-multipoint call/connection allows the distribution of an uni-directional flow of data from one source, the root party, to a set of sinks, the leaf parties. The recommendations Q.2971 at the UNI [2] and Q.2721.1 at the NNI [3] specify the point-to-multipoint calls. These calls are always initiated by the root party, which can add and remove further leaf parties. A leaf party can only drop itself from the call, but cannot join itself to the call. The advantage of point-to-multipoint calls is that the network can perform ATM multicasting in the appropriate node, which can result in a large bandwidth saving.

Traditional IN services, like number translation services, reverse charging services, VPN services could be extended to point-to-multipoint. For example a one-to-many number translation feature could allow a single called number to trigger IN service logic which could add a set of leaves to the point-to-multipoint call. A large set of services based on the Intelligent Broadband concepts could benefit from the adoption of the point-to-multipoint. On one hand, services like B-VC could be realized in a more efficient way by using point-to-multipoint connections. On the other hand, distributive services with centralized control could be realized (e.g. TV-distribution or Distance learning services).

The call and session model must be extended to describe the evolution of the call and of the different parties in a point-to-multipoint call. The extended model should allow the control of the basic procedures of a point-to-multipoint call.

Moreover to be compliant with the Intelligent Broadband architecture, the model must take into account some special features, like the addition of parties on behalf of the IN and the SCP-initiated point-to-multipoint call.

In a point-to-multipoint call there is a set of parties which can evolve independently, making it not feasible to control them within a single state machine. A possible solution [4] is to use the BCSM defined for point-to-point call in Section 2.2.4 to handle the aspects related to the entire call and to add a set of state machines representing each remote party. In the following these state machines will be called 'Call Control Manager' (CCM) and 'Party Control Manager' (PCM). The structure of the PCM is the same as the CCM, only the names of PICs and DPs are changed. Figure 3.3.1 shows the first PICs of Party Control Manager. It is an easy exercise to derive the full model starting from the Call model shown in Section 2.2.4.

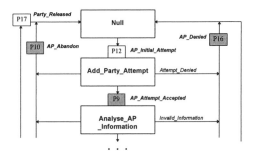

**Figure 3.3.1**
Party Control Manager state machine

In a 'basic' point-to-multipoint call (i.e. setup by the root), a PCM is initiated when the setup indication is received from the root, and will evolve in parallel with a CCM which is initiated as in the point-to-point case. The CCM represents the call state while the PCM represents the state of the first remote party. The remote parties that will be successively connected will be monitored by new instances of Party Control Manager, as shown in Figure 3.3.2. Note that only the control of the originating side of the call is envisaged in this description, but the model can be extended to include also the terminating side aspects.

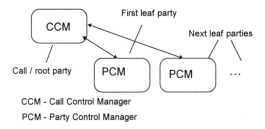

**Figure 3.3.2**
BCSM instances for the control of a point-to-multipoint call

For the addition of parties on behalf of the IN and the SCP-initiated point-to-multipoint calls, it is possible to follow the same approach as for SCP-initiated calls. These features will be supported via enhancements of the call handling functionality, without modification of the signaling protocols. For example to realize the SCP-initiated point-to-multipoint call, a point-to-point call is first established to support a unidirectional connection from the root to the node, then an independent (with respect to B-ISDN signaling) point-to-multipoint call towards the leaf parties is setup. The modified call handler will correlate the signaling messages, will dialogue with IN service logic, and will control the through-connection of the user plane links. In the model, the leaf parties will be represented by PCM instances, with the usual behavior. In this case the CCM can control the root party. The CCM will pass in the active state when the root has answered the call, therefore it could be allowed to have an active link towards the root with no leaf party connected. In a broadcast service this could be useful for having data flow always ready to be distributed to users requiring it.

The session model needs to be slightly enhanced to support the point-to-multipoint calls. The *Bearer Connection* object can be extended to support a set of more than two *Legs*. A new attribute '*Configuration*' will be added to the bearer connection object with possible values '*point-to-point*' and '*point-to-multipoint*'. *Leg* objects will represent the root and the set of leaves which compose a point-to-multipoint call. The set of interaction diagrams for the support of point-to-multipoint could be extended with two new diagrams as reported in Table 3.3.1. The meaning of the interaction diagrams is straightforward.

**Table 3.3.1**
New SSF-SCF interaction diagrams for the support of point-to-multipoint

| Interaction diagrams | Direction |
|---|---|
| Add point-to-multipoint bearer | SCF → SSF |
| Join parties to point-to-multipoint bearer | SCF → SSF |

## 3.3.2   Architectural Enhancements for Multi-connection Calls

The call and connection control separation is an outstanding feature for the B-ISDN, its main purpose is to avoid waste of resources and to allow a more powerful control of the transport resources. The multi-connection calls represent a very important consequence of the call and connection control separation. A *call* will represent only the end-to-end association between users and conceptually has no user plane resources associated. A set of *connections* can be associated with the call and will represent the user plane links. A general model for the multi-connection calls in the Intelligent Broadband Network will be analyzed hereafter although it has to be noted that there is no stable protocol specification for this mechanism.

As in the point-to-multipoint call, the main requirement is to control a set of independent concurrent events at the call level and at the level of the different connections. The most effective and modular solution is to define a two-level model including a single state machine at the call level and a set of state machines at the connection level. These machines will be called Call Control Manager (CCM) and Bearer Connection Control Manager (BCCM). The CCM could be identical to the BCSM defined in Section 2.2.4 and its task is to monitor and control the call related aspects: collecting the requests from the originating user, checking the authority/capability of the user, establishing the call relationship towards the called user. The BCCM is a simpler state machine used to monitor and control the events related to a single connection: (e.g. the establishment and the release of the user plane link). Figure 3.3.3 shows an instance view of the two-level call model for multi-connection calls.

It is worth noting that the introduction of multi-connection calls could be even hidden to the service logic. It is possible to modify only the mapping of session level operations into CCF operations, to take into account the more powerful capability of the B-ISDN network. This mapping is performed within the SSF, so that the session model and the SSF-SCF interface will not be affected.

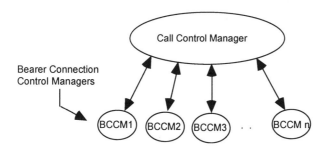

**Figure 3.3.3**
BCSM instances for the control of a multi-connection call

# References

[1]  Ronchetti L, Paglialunga A, Siviero M, Faglia L and Minzer S, *A resources reservation protocol for B-ISDN services*, Proceedings of IEEE International Telecommunication Symposium, Rio de Janeiro, August 1994
[2]  ITU-T Recommendation Q.2971, *Broadband Integrated Services Digital Network (B-ISDN) – Digital Subscriber Signalling System No. 2 (DSS 2) – User-Network Interface Layer 3 Specification for Point-to-Multipoint Call/Connection Control*, Geneva 1995
[3]  ITU-T Recommendation Q.2721.1, *B-ISDN User Part – Network Node Interface specification for point-to-multipoint call/connection control*, Geneva 1996
[4]  Listanti M, Salsano S, *Point-to-multipoint call modeling for IN/B-ISDN integration*, Proceedings of 2IN'97, Paris, September 1997

# Chapter 3.4

# RELATIONSHIP OF INTELLIGENT BROADBAND NETWORKS WITH NARROWBAND INTELLIGENT NETWORKS

One objective of the Intelligent Broadband Network concept considered in this book is to ensure a smooth migration from narrowband towards Broadband. A verification of this objective can be performed by comparing the Broadband IN concepts with the concepts of narrowband IN, and by exploring the coexistence of both concepts within one network. Such a coexistence can be achieved by an interworking of narrowband and Broadband network elements.

Since many SCPs are already deployed within the narrowband ISDN, and many IN services are executed successfully by them (e.g. Freephone, Premium Rate and also Virtual Private Network [1]), it is reasonable to consider an interworking of such narrowband SCPs (N-SCPs) with B-SSPs. Such an interworking would allow one to use existing narrowband SCPs and services within a B-ISDN. For the lower protocols of the communication stack between B-SSP and N-SCP (including the TCAP layer), the usual interworking mechanisms between an ISDN and an ATM network can be applied by replacing the MTP layers for narrowband signaling by the SAAL and MTP-3b layers for Broadband signaling (see [2], Chapter 1.7, or [3], Chapter 3.3).

More interesting is the interworking at the INAP level, because, at this level, the main semantic differences of the messages appear. The following chapters show examples of the interworking between B-SSPs and N-SCPs at the INAP level.

Technically, there are two ways of interworking between an N-SCP with a B-SSP. One way would be to use the N-INAP for the communication between the N-SCP and the B-SSP and to execute the N-INAP messages within the B-SSP. However, this requires a modification of the B-SSP software. Examples for this variant are discussed in Section 3.4.1 in more detail. The other way is to place an Interworking Function (IWF) between the B-SSF and the N-SCF which translates B-INAP messages into N-INAP messages and vice versa.

This interworking functionality is not difficult to achieve, since the B-INAP described in Chapter 2.3 is designed already in a way which allows a certain form of backwards compatibility with the CS-1 or CS-2 INAP (N-INAP) for N-ISDN.

The advantage of such an IWF is that it can be placed either within the B-SSP, or within the N-SCP, or even within a separate network element. In the latter

*Intelligent Broadband Networks.* Edited by I. S. Venieris, H. Hussmann
© 1998 John Wiley & Sons Ltd.

case, no modifications at the B-SSP and the N-SCP are required. Further information about such an IWF is given in Section 3.4.2.

## 3.4.1   Interworking by using the N-INAP

In the case where the communication between a B-SSP and an N-SCP uses the N-INAP [4, 5], the B-SSP must be modified so that it can execute the N-INAP operations. Figure 3.4.1 shows an architecture which allows the communication of a B-SSP with both a B-SCP and an N-SCP.

The execution of the N-INAP operation is performed by the functional block N-SSF, which can be similar to the SSF within an N-SSP. However, the specialties of the B-CCF compared with the N-CCF of an N-SSP must be taken into account.

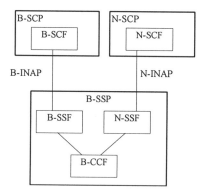

**Figure 3.4.1**
Functional Model of a B-SSP supporting the connection of both B-SCPs and N-SCPs

For the realization of such a B-SSP, however, it is not necessary to have completely separate software for the B-SSF and the N-SSF. For example, the mechanisms used at the interface between CCF and SSF and at the interface between SSF and protocol stack handling (TCAP) can be used for both SSF-variants. Figure 3.4.2 shows an extension of the software of the Siemens B-SSP described in Chapter 2.5, realizing the N-SSF.

Since the N-INAP messages operate directly on the (B-)CCF, no translation between SSM and BCSM concepts is needed. Therefore, the narrowband IN Switching Manager, performing the SSM handling, will be very simple compared with the handling of the B-SSM. The functionality of the narrowband SCP Access Manager is the same as the functionality of the (Broadband) SCP Access Manager, except that N-INAP operations instead of B-INAP messages are extracted from or inserted into TCAP-messages and forwarded to or received from the SSM handling.

A prototype for the interworking of an N-SCP with a B-SSP is described in [6].

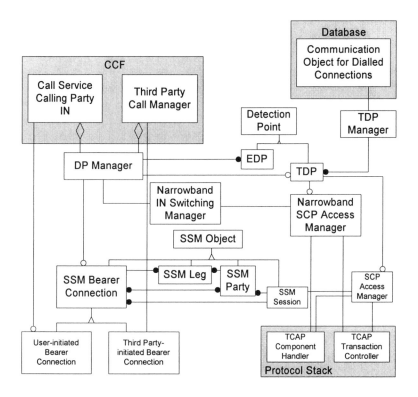

**Figure 3.4.2**
Extended Class Diagram for the N-SSF Implementation

## 3.4.2   Interworking by using an Interworking Function (IWF)

Since one of the goals of the Intelligent Broadband Network is to ensure a smooth evolution from existing narrowband IN towards Broadband IN, a possible interworking with narrowband has been kept in mind during the development of the respective architecture. The powerful session concept, introduced in the Intelligent Broadband Network architecture makes it compatible with the architecture of narrowband IN: since narrowband IN operates only on single connections, it can be assumed that narrowband IN handles very simple  sessions with exactly one bearer connection. This implies that, at an abstract level, the architectural model for narrowband IN is a subset of the Intelligent Broadband Network model.

A second architectural aspect important for interworking is the modeling of the view on the network provided to the SCP: In narrowband, the BCSM provides this view [7, 8], whereas for Intelligent Broadband Network, the IN-SSM introduces a more abstract view than the BCSM. However, states of the BCSM can be easily mapped to IN-SSM states. Doing so, mapping between the different INAP messages operating on these different models is necessary.

To perform a mapping of N-INAP operations, which are used by an N-SCP, to B-INAP, used by a B-SSP, an Interworking Function (IWF) must be provided between the B-SSF and the N-SCF, as shown in Figure 3.4.3.

As mentioned before, the IWF can be located either within the B-SSP, or within the N-SCP, or within a separate network element.

In principle, there are two possibilities to realize a dialogue between the B-SSP and the N-SCP. Either there are two separate TCAP-dialogues between B-SSF and IWF and between IWF and N-SCF, or there is only one TCAP-dialogue between B-SSF and N-SCF. In the latter case, the IWF only handles the INAP-level of the protocol stack and leave the other lower levels unchanged. The first possibility is preferred if the IWF is located within a separate network element, whereas the second possibility saves protocol handling overhead if the IWF is located either within the B-SSF or the N-SCF. The selection of either possibility should be transparent to both the B-SSF and the N-SCF.

The interaction diagram shown in Figure 3.4.4 gives an example of the translation between B-INAP and N-INAP performed by the IWF.

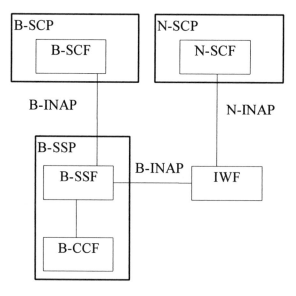

**Figure 3.4.3**
Functional Model containing the Interworking Function

In this example, the B-SSP requests a simple number translation service, which is performed by the Service Logic of an N-SCP. The Service Logic for such a request is usually invoked by the InitialDP message. Furthermore, this Service Logic wants to be informed of the success of the establishment of the call by sending a RequestReportBCSMEvent message. The translated number is contained in the Connect message from the N-SCP, whereas the report of the successfully established call is received within the ReportBCSMEvent message.

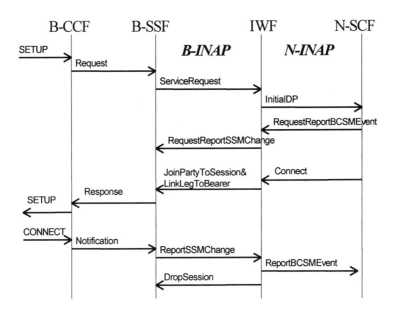

**Figure 3.4.4**
Interaction diagram showing the activities of the Interworking Function

This diagram shows that a pure translation of messages is not sufficient: For the implementation performed in the context of the INSIGNIA project (see Chapter 2.8), an SCP-SSP relationship is always (except error cases) explicitly terminated by the SCP, whereas in IN CS-1/CS-2, an SCP-SSP can also be terminated by the SSP, namely in the case where no further BCSM event is requested to be reported. Therefore, the IWF must monitor pending BCSM events to be reported, and must eventually terminate the session within the B-SSP by e.g. DropSession.

# References

[1]  Thörner J, *Intelligent Networks*, Artech House, Boston/London, 1994
[2]  Killat U, *Access to B-ISDN via PONs*, Wiley/Teubner, Chichester/Stuttgart, 1996
[3]  Chen T and Liu S, *ATM Switching Systems*, Artech House, Boston/London, 1995
[4]  ITU-T Recommendation Q.1218, *Interface Recommendation for Intelligent Network CS-1*, Helsinki, 1993
[5]  ITU-T Recommendation Q.1228, *Interface Recommendation for Intelligent Network CS-2*, Geneva, 1997
[6]  Van der Vekens A W, *An Object-Oriented Implementation of B-ISDN Signalling - Part 2: Extendibility Stands the Test*, Proceedings of 18th International Conference on Software Engineering, Berlin, March 1996
[7]  ITU-T Recommendation Q.1214, *Distributed Functional Plane for Intelligent Network CS-1*, Helsinki, 1993
[8]  ITU-T Recommendation Q.1224, *Distributed Functional Plane for Intelligent Network CS-2*, Geneva, 1997

# Chapter 3.5

# RELATIONSHIP OF INTELLIGENT BROADBAND NETWORKS WITH INTERNET

Public telecommunication networks including the Broadband ISDN have been developed out of a completely different tradition than the Internet and its services. The Internet was originally a computer network dedicated to a special group of users (military and academic, mainly) but during the last decade it has evolved into a publicly available infrastructure which is, in particular, adequate for bringing multimedia services to a large group of users. Sometimes this is considered a general deep conflict between the 'information technology' world and the 'telecommunications' world, and a competition is seen between these two worlds for the new multimedia services which will dominate the future of telecommunication. However, the developments of the last few years have shown that the two worlds are converging. There are some attempts to base traditional telecommunications services (e.g. phone and fax) on Internet technology, and on the other hand a large amount of the Internet traffic is carried over traditional public telecommunication networks like the ISDN or public ATM networks.

This chapter tries to explain how a coexistence and synergy between the Internet and the telecommunication concepts can be achieved, using the Intelligent Broadband Network as a representative for the telecommunication network side. It starts with a general top-down analysis of the potential ways how Intelligent Broadband Networks and Internet can be brought into contact. As a concrete example, one concept of particular relevance is discussed in more detail, which is the idea of a 'virtual intranet'.

## 3.5.1 Combining Intelligent Broadband Networks with Internet

The Internet can be seen from several different viewpoints. First of all, the Internet constitutes a specific technology for building and running computer networks. The core part of this technology is a layered set of communication protocols based on the common Internet Protocol [1]. This technology was designed to be extremely flexible. In particular it keeps large parts of the network services independent from the underlying link layer and physical network, and therefore it enables large heterogeneous networks. Moreover, the protocols are able to cope with a changing network topology and realize a high degree of self-organization of the network. Many services are available on the Internet besides the extremely popular World Wide Web (WWW).

*Intelligent Broadband Networks*. Edited by I. S. Venieris, H. Hussmann
© 1998 John Wiley & Sons Ltd.

Apart from this technology viewpoint, 'the Internet' now is a rapidly growing global network of high capacity with a completely different business model than traditional telecommunication networks, where e.g. the cost of communication is independent of actual used bandwidth usage and geographic distance.

When discussing a combination between Intelligent Broadband Networks and Internet, it therefore makes sense to distinguish between the *usage of Internet technology* in an Intelligent Broadband Network and the *combination of two physically separate networks*.

## 3.5.1.1 Usage of Internet Technology in an Intelligent Broadband Network

In the case of usage of Internet technology, the global Internet infrastructure is not involved at all. Instead, some of the protocols constituting the Internet technology are used to construct services of the Intelligent Broadband Network. This means that on top of some ATM virtual connections the Internet Protocol is used, enabling further protocols and services from the Internet world, e.g. reliable information transport using the Transmission Control Protocol (TCP), or the WWW service using the Hypertext Transfer protocol (HTTP).

This approach is particularly interesting for the development of applications running on end systems since support for the Internet protocol suite is readily available for any hardware and software platform. For example, an application realizing a Video on Demand service to the user can be based on a WWW service which makes it very easy to provide a simple and reliable user interface. In the experiments performed in the context of the project INSIGNIA (see Chapter 2.8), this approach was chosen for the demonstrated Video on Demand service. It is worth noting that the Intelligent Broadband Network here still is responsible for building up the communication configuration. The Internet protocols are just used inside the ATM virtual connections and can benefit from the fact that the Broadband network reserves adequate bandwidth to ensure high performance.

Another example where it is logical to use Internet technology is the communication between an end system and a Broadband Intelligent Peripheral. In most cases, user dialogues have to be carried out here which can be realized using the WWW service, and possibly enhancing these services with platform-independent applets written in the Java language. The advantage of the Internet technology is here that the same kind of interface works for a large variety of different end systems.

To summarize, the usage of Internet technology can achieve a high degree of synergy in those parts where end systems are attached to the Intelligent Broadband Network.

## 3.5.1.2 Combining Heterogeneous Networks

When considering the physical combination of an Intelligent Broadband Network (IBN) with the Internet, again several cases can be distinguished which are summarized in Figure 3.5.1.

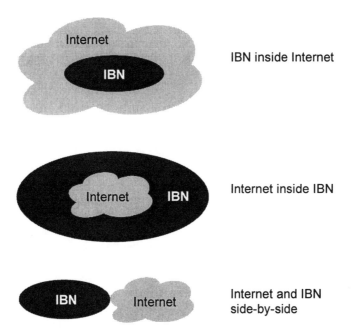

**Figure 3.5.1**
Combining Intelligent Broadband Networks and the Internet

The first alternative is *IBN inside Internet* which means that the user of such a combined network has the impression to work with the ordinary Internet. The only difference is that some services between some nodes of the network can be enhanced in terms of capacity, security, or business model. Of course this means that the Intelligent Broadband Network is used for a different kind of services than the multimedia services described in other parts of this book. However, the network-oriented services (for which also an example is covered in this book) can be used very well for enhancing the internal infrastructure of the Internet. One example of this kind of combination is described in more detail below (the Virtual Intranet). Another variant of this approach is to provide a special gateway by which some Internet servers are able to influence the service control function of the Intelligent Broadband Network Using such a gateway, an Internet server may for instance instruct the Broadband network to establish a particular connection with a particular billing scheme. Complex service logic may be required for the optimization of such requests, e.g. for combining such individual requests to achieve an optimal use of resources. An example for a prototype system following this idea can be found in [2]. The system described there is based on TINA/PSTN instead of IN/ATM but the ideas can be transferred also for Intelligent Broadband Networks.

The second alternative is *Internet inside IBN* such that the Intelligent Broadband Network used some Internet internally. An example application is a collection of Intelligent Broadband Network islands which are spread geographically and which lack ATM interconnectivity between the islands. The

Internet can be used to interconnect Service Control Points of the islands, and therefore enable some level of interworking among services in the various islands. This type of combination seems to be less important than the first one.

The third alternative, finally, is to put *Internet and IBN side-by-side*. This approach leaves both networks unchanged but interconnects them, using some kind of gateway function. The classical application is here to use an Intelligent Broadband Network as a kind of high-speed access network to an Internet-structured network, in a similar style as ISDN and PSTN are nowadays used to access the Internet. This does not yet make much sense in the current situation of the global Internet where only low bandwidth can be provided for the individual user. But this approach can be very interesting for connecting users to a high-speed network which is built according to the Internet principles but physically separated from the Internet. Such a network may be for example an interconnection of high-performance database, media or computing servers or a corporate network with high bandwidth. The (public) Intelligent Broadband Network in this case can be used to provide functions which are well known from narrowband Intelligent Networks. An example is routing to an appropriate gateway dependent on origin and/or time of day. Moreover, the Intelligent Broadband Network can improve security since it knows about the physical access point to the network. Finally, it can apply special charging schemes for the ATM virtual connections, e.g. based on membership in some organization.

The next part of this chapter concentrates on one specific way to combine Intelligent Broadband Networks and the Internet as physical networks. It is a special case of the second alternative from above (IBN inside Internet).

## 3.5.2 Broadband Virtual Intranet

### 3.5.2.1 Virtual Intranet

Many large enterprises today are using computer support to enhance their traditional infrastructure for information retrieval and business processes. It has turned out that the concepts of the Internet and in particular of the World Wide Web are able to solve many of the problems which come from the heterogeneous and distributed way in which information is available in such organizations. Therefore, large companies build up internal networks which use Internet concepts but are separated from the world-wide Internet through 'firewalls'. Such networks are called *Intranets*.

However, most Intranets today rely on a physical separation from the outside world. This is a safe solution but not adequate for the modern style of working in global organizations where people are traveling or moving around rather frequently and would like to continue their work in the same way from many different locations. Nevertheless, the organization is interested in keeping this traffic clearly separated from other traffic on the Internet. So the physical separation of the Intranet is made virtual, which leads to the notion of a *Virtual Intranet*.

There are several technical solutions for the realization of a Virtual Intranet. One class of solutions is based completely on the techniques of the Internet and its protocol stack (Internet Protocol, in particular the new version 6). In these approaches, users of the Virtual Intranet are expected to connect to the public Internet. Encryption schemes are used to ensure that the traffic of the Intranet crossing the public sections of the Internet is kept confidential. However, this approach has some drawbacks. Generally, it is considered dangerous by most organizations to encourage corporate users to connect to the public Internet. It is also not yet clear how adequate billing for such Intranet traffic on the Internet can be achieved. Moreover, the public Internet as it is today does not provide the required bandwidth for broadband applications like Computer-Supported Co-operative Work (CSCW).

Therefore, alternative solutions are still attractive for corporate networks. We explain below a way how the features of a Broadband Intelligent Network, in particular a Broadband Virtual Private Network service, can be used to provide a potentially superior solution.

## 3.5.2.2 Virtual Intranet based on an Intelligent Broadband Network

The basic idea of the Intelligent Broadband Network approach to Virtual Intranets is again to keep the Intranet users separated from the Internet. However, this separation is not achieved on the physical level but on a logical level which is located between the physical network and the Internet protocols. The B-ISDN is used to provide the physical (high-bandwidth) connectivity of terminals, and the Broadband Intelligent Network enables the required functionality of virtual networks.

Figure 3.5.2 shows an example configuration for such a Virtual Intranet. The typical usage of such a configuration will be that the physically external Intranet users access WWW servers in the Intranet or establish CSCW sessions with other Intranet users. Of course, the Intelligent Broadband Network can also interconnect separate LAN islands of the Intranet.

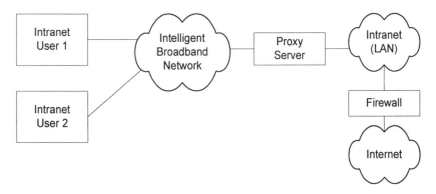

**Figure 3.5.2**
General configuration of a Broadband Virtual Intranet

In the configuration of Figure 3.5.2, the Intelligent Network is used to provide the functionality of a Broadband Virtual Private Network (B-VPN, see Section 2.4.1). The B-VPN service enables the owner of the Virtual Intranet to administer his own range of ATM addresses (IN numbers) and to define specific privileges for calls established to or from these numbers. Moreover, special conditions for billing can be arranged with the network operator for these numbers.

In order to make Internet applications (using the Internet Protocol stack) work over an ATM network, some specific encapsulation and address resolution mechanisms are required. Fortunately, these mechanisms are specified already by Internet standards. RFC (Request for Comment) 1483 and RFC 1577 define the so-called 'Classical Internet Protocol over ATM' [3, 4]. Implementations of these standards are already available from several manufacturers. According to Classical Internet Protocol over ATM, a special address resolution server (ATM ARP server) has to be made available to the ATM end systems. This server essentially holds a translation table from Internet Protocol addresses to ATM addresses. The drivers for Internet Protocol over ATM take care of the establishment of ATM switched connections towards the respective ATM addresses as soon as Internet Protocol traffic towards a new Internet Protocol address appears. Connections are also torn down automatically after some defined idle time.

For the Broadband Virtual Intranet, such an ATM ARP server is required. However, the ATM (E.164) addresses used in the tables of this server should have the format of IN numbers denoting members of a B-VPN. This way, it is automatically achieved that the Internet Protocol over ATM driver, when establishing an ATM connection, uses an IN number and so invokes the B-VPN service. Additional security is achieved if all ATM end systems of the Intranet (external users and proxy servers) refuse any incoming calls which were not processed by the B-VPN.

The effect of such a configuration is that the Intranet users usually do not realize that they are working with the Intelligent Broadband Network. However, the Internet Protocol over ATM mechanism automatically ensures that IN calls are established which means that the security and integrity of the VPN can be controlled by the IN service logic. Moreover, IN enables the usage of specific billing for VPN calls, for example a mass discount scheme. Another advantage is that an Intranet user can stay in the same subnetwork of the Intranet (the logical subnetwork formed by Internet Protocol over ATM) even if the physical location of a user or computer changes.

To summarize, the IN service B-VPN can be used effectively for enhancing Intranets towards Broadband Virtual Intranets. This proves that Internet technology and Intelligent Broadband Networks are not competitors, but can be used in an integrated and synergetic way.

# References

[1]   Comer D, *Internetworking with TCP/IP*, Prentice-Hall 1989

[2]  De Zen G, Marsiglia M A, Ricagni G, Vezzoli G, Hussmann H, Schoenbauer H, Sevcik M, Zoernack A, *Value-Added Internet: a Pragmatic TINA-Based Path to the Internet and PSTN Integration*, Proceedings of TINA'97 Conference, Santiago de Chile, 1997

[3]  Heinanen J, *Multiprotocol encapsulation over ATM adaptation layer 5*, Internet RFC 1483, Telecom Finland, July 1993

[4]  Laubach M, *Classical IP and ARP over ATM*, Internet RFC 1577, Hewlett-Packard Laboratories, January 1994

# Part 4

## PERFORMANCE DRIVEN DESIGN OF INTELLIGENT BROADBAND NETWORKS

# Chapter 4.1

# PERFORMANCE ISSUES IN INTELLIGENT BROADBAND NETWORKS

The common practice followed for reducing the risks entailed to the introduction of any novel system in terms of cost and expected performance, is to assess, prior to the development phase, the system behavior under various conditions. Having ensured that the system operates according to the specifications one has to consider ways for enhancing the system performance.

Performance evaluation constitutes a widely accepted technique for estimating the system response to various excitations as well as for obtaining both qualitative and quantitative results that provide valuable feedback to the system design process. In particular for signaling which constitutes the neural system of a network, performance evaluation has always been a tentative engineering task as it determines to a high degree the final system to be deployed. This was the situation for Signaling System No 7 (SS7) and still is for B-ISDN signaling as becomes evident from a high number of relevant publications that investigate the various issues which affect the performance of a call handling system [1-9]. In particular for Intelligent Broadband Networks, performance issues become more evolving due to the integrated service facilities offered by the network, the requirement for frequent user-network interactions and the employment of IN capabilities across the network.

## 4.1.1 Study Items

The introduction of the Intelligent Network (IN) paradigm in conjunction with a broadband infrastructure represented by the B-ISDN allows a flexible provisioning of advanced multimedia services. A service is created in a modular way by using Service Independent Building Blocks; this flexibility allows network operators to upgrade their systems easily and in a cost effective manner by taking advantage of the ability to allocate, without restrictions, control functionality and resources within the network.

In Intelligent Broadband Networks, services are described by a set of distinctive characteristics that are implemented by various functional entities distributed across several network elements. The itinerary an IN service call will follow involves several functional entities located at the same or remote physical entities of the network (see for example the interaction diagram of the services described in Chapter 2.4). A functional entity may be visited by a request more than one time as the process of call establishment may require extensive

*Intelligent Broadband Networks*. Edited by I. S. Venieris, H. Hussmann
© 1998 John Wiley & Sons Ltd.

transactions among the involved functional entities. Moreover IN services require extended processing in the respective functional entities not only during the set up and release phase of a call but also throughout the call lifetime. In this respect, the load of a Broadband signaling system is expected to be considerably heavier than of single service networks or N-ISDN and the possibility of performance degradation becomes more vivid. If so, user requests for services will be denied or at best the delay experienced for processing a call will not be acceptable to the user who most probably will reattempt to call or will abandon the service.

All these newly introduced parameters force to the development of complex user-to-network and network-to-network transaction models able to capture the effect of sophisticated protocol layers, signaling information databases and service control architectures. It can be easily understood that the performance evaluation of Intelligent Broadband Networks is not a trivial task.

One of the most important issues is the methodology used to model the flow of messages accompanying the establishment of an IN call. This should capture realistic scenarios regarding the user behavior while being accurate at the protocol level where messages and actions are strictly defined by the relevant specifications. In Chapter 4.2 the traffic modeling approaches followed in this book are presented along with a set of models for accurately representing the functional entities and signaling protocols involved in the processing of a call request. These tools have been developed to enable the evaluation of a number of performance issues related to the design and operation of an Intelligent Broadband Network. Such studies appear in the subsequent chapters of this part. Before presenting the methods and the respective models, the performance issues addressed in this book are briefly outlined.

It is widely accepted that any network should be able to accept the 'maximum' number of user requests using as criteria the availability of the requested resources (e.g. bandwidth, buffers), and the user-requested Quality of Service. Hence it is important for a signaling system to guarantee that at least those calls that can be accommodated by the network can be processed with acceptable delay. Furthermore the signaling network should have a scalable architecture and topology able to accommodate the future growth in service requests.

From a performance perspective, the most significant aspect introduced by the IN paradigm is the realization and the distribution of the network intelligence that, in strict cooperation with the B-ISDN, realizes the service delivery. The performance evaluation is focused on the derivation of design criteria to structure the 'intelligent level' of the network. This issue is faced by taking into account that the emerging services require intense interactions with the network, resulting in the exchange of a high number of control messages. As a consequence the distribution of IN functionality into different architectural structures of the network (i.e. the 'functionality mapping' issue) becomes a dominant factor in the overall system performance [10, 11].

In tele-traffic engineering terms the above requirement maps onto the problem of a balanced sharing of the load at all levels of the Intelligent Broadband Network, that is the physical entities as well as the protocol layers and the functional entities within each physical entity. More precisely at the Intelligent

Broadband Network level one has to choose the most effective way for allocating functional entities to physical entities while satisfying scalability, throughput and call set up delay requirements. Attempting an anatomy to a single physical entity, one should then decide on the way the several functional entities and functions within the protocol layers are serviced. Therefore single processor vs. multiple processor architectures at both the physical entity or protocol level should be investigated as well as priority schemes for scheduling the execution of different tasks with a protocol layer.

Since the service logic is distributed among different network elements, IN services can shift the performance bottlenecks from switching nodes to other physical entities. As a result, efficient congestion and overload control mechanisms, tailored for the specific IN environment, must be studied in order to guarantee an appropriate Quality of Service. The action of congestion control under overload conditions is necessary and inevitable in order to prevent network failures. It aims to maximize throughput (demonstrating, at the same time, a good delay performance) by adjusting the offered load. If no appropriate actions are taken, message loss will result in poor performance. For optimal utilization of network resources, a service request should not be accepted if the network can not support it. The utilization level of functions in combination with the processing capacity could give a criterion for the acceptance of the service.

# References

[1]  Kuehn P, Pack C, Skoog R, *Common Channel Signaling Networks: Past, Present, Future*, IEEE J. Select. Areas Commun., **12**, 383-394, 1994

[2]  Bafutto M, Kuehn P, Willman G, *Capacity and Performance Analysis of Signaling Networks in Multivendor Environments*, IEEE J. Select. Areas Commun, **12**, 490-500, 1994

[3]  Lazar A, Tseng K, Lim K, Choe W, *A Scalable and Reusable Emulator for Evaluating the Performance of SS7 Networks*, IEEE J. Select. Areas Commun., **12**, 395-404, 1994

[4]  Hou X, Kalogeropoulos N, Lekkou M, Niemegeers I, Venieris I, *A Methodological Approach to B-ISDN Signaling Performance*, Int. J. of Commun. Sys., Special Issue: Signaling Protocols and Services for Broadband ATM Networks, 7, 97-111, 1994

[5]  Lekkou M, Venieris I, *A Workload Model for Performance Evaluation of Multimedia Signaling Systems*, Comp. Commun., **20**, 884-898, 1997

[6]  La Porta T, Veeraraghavan M, *Evaluation of Broadband UNI Signaling Protocol Techniques*, Journal of High Speed Networks, **2**, 209-238, 1993

[8]  Ghosal D, Lakshman T V, Huang Y, *Parallel Architectures for Processing High Speed Network Signaling Protocols*, IEEE/ACM Trans. Netw., **3**, 716-728, 1995

[9]  Hwang R-H, Kurose J, Towsley D, On call Processing Delay in High Speed Networks, IEEE/ACM Trans. Netw., **3**, 628-639, 1995

[10] Kolyvas G, Polykalas S, Venieris I, *A study of specialized resource function mapping alternatives for Integrated IN/B-ISDN architecture*, Proceedings of 2IN'97, Paris, September 1997

[11] Cuomo F, Listanti M and Pozzi F, *Provision of broadband video conference via IN and B-ISDN integration: architectural and modeling issue*, Proceedings of 2IN'97, Paris, September 1997

# Chapter 4.2

# METHODS AND MODELS FOR PERFORMANCE ANALYSIS

This chapter attempts an overview of the existing approaches for modeling signaling and service control systems and evaluates their applicability in the context of Integrated Broadband Networks. A concrete methodology is established for modeling the system entities considering both a high level representation of the system focusing at the network element as a whole and a low level one for modeling in detail protocols and processes. Particular emphasis is put on the generation of the workload for these models which in general should reflect the particular type and sequence of messages exchanged internally between the processes of a protocol as well as between protocols and between physical entities. The chapter starts the presentation of the models by discussing state of the art techniques used to model signaling traffic. It then continues with IN traffic models and closes with the presentation of a set of protocol and functional entity models which are used in the performance studies of the subsequent chapters.

## 4.2.1 Traffic Models

To study the performance of a telecommunication system in general and a signaling system in particular, a set of models should be developed for accurately representing (signaling) traffic flows. Traffic modeling is a key issue in performance evaluation, and significant research effort has been allocated on the development of accurate and easy to handle traffic models. Regarding signaling interaction diagrams two main trends can be identified in the literature: The first assumes that signaling message flows follow distributions with well known properties, whereas the second uses pre-determined scenarios of signaling messages. A review of the current approaches in modeling signaling traffic follows.

### 4.2.1.1 Review of Existing Approaches

Kosal, and Skoog [1] assume that signaling message arrivals is a Poisson process on account of the fact that call arrivals process is Poisson. The SS7 offers to users the ability to communicate end-to-end using signaling messages. Hence, users can exchange large data blocks which are segmented into consecutive signaling messages. Despite the fact that these messages are correlated, the interarrival times can be assumed independent when the call arrival process is Poisson. This can be justified by considering that the interarrival time between consecutive messages is large enough compared to the length of the most busy periods. In this case a correlated arrival comes with high probability after a busy period and the system has lost all memory of the previous events. Hence, the system behaves as though the arrivals were not correlated. Different types of signaling messages are used, and an average length of each message type is used for the performance evaluation.

Manfield *et al.* [2] classify the signaling messages in message types and assumes that the message arrivals for each class is a Poisson process. The message length distribution depends on the higher layer and is different for each class. Also in each distribution there are messages with variable length because of their specific parameters. However a default value is used for the message length.

In contradiction with the above studies which assume Poisson arrivals for the signaling messages Duffy *et al.* [3] have pointed out that even if calls arrivals follow a Poisson process, message arrivals will not follow a Poisson distribution because they have been correlated in a particular call. This result was derived by statistical analysis of SS7 messages which were monitored in a Signaling Transfer Point.

Zepf, and Rufa [4] suggest three types of signaling sources. The first assumes Poisson arrivals. The second consists of several signaling scenarios that are triggered by a simple generator. The generator is characterized by a given distribution for the interarrival time and the mixing of signaling scenarios. Each signaling scenario is described by the number of signaling messages, their individual priority and by the distribution functions for the interarrival times. If the first message of the set up phase is accepted, subsequent signaling messages belonging to the same scenario should not be affected by any load reduction scheme. This can be achieved if the first message is assigned the lowest priority. In case the first message of a signaling scenario is rejected, the subsequent messages of the scenario will be suppressed by the generator. The third model introduces the call re-attempts when the call fails. The reattempt probability, the maximum number of re-attempts and the distribution of the time between the call rejection and the reattempt are deterministic.

Smith [5] assumes that signaling messages follow a simplified scenario, which is triggered when a call is originated. This scenario is based on the assumption that a call is always answered and that the calling part always disconnects first. It also assumes that no continuity test messages are transmitted. The signaling scenario is invoked whenever a call request appears. In this work the call arrivals follow a Poisson process, messages are generated with constant delays, the holding time is based on the work of Bolotin [6] that describes real calls. When the call fails the calling part re-attempts by Bernoulli trials.

Rumsewicz [7] approaches the problem in a similar way with Smith [5] but without the assumptions that every call is answered and that the calling part always disconnects first. The ringing time follows a uniform distribution, the holding time is based on the work of Bolotin [6] and the re-attempt time follows a uniform distribution.

A more realistic case is investigated in [8] by Baffuto *et al.*, where signaling messages are organized in various scenarios representing the mix of services. For each service supported by the network more than one sub-scenario may exist corresponding to the possibility of successful call, absence of the called user etc. The load information is computed on a per sub-scenario basis and each sub-scenario is assigned a constant probability.

Bolotin [6] shows that the exponential distribution does not describe adequately the holding time of a call, but instead the most suitable distribution is a composition of lognormals. Since holding time affects the signaling traffic - the total number of queries depends on the call duration - the selection of the holding time distribution has especially importance for the signaling traffic.

## 4.2.1.2 Intelligent Network Traffic Models

Although, traffic models have been widely investigated for SS7 based networks, analogous issues have not been examined in the context of Intelligent Broadband Networks. For simple signaling protocols that support a restricted number of well known services the information flows can be easily defined. The IN capability of the signaling network allows for services with very different characteristics in terms of protocol and functional entities interactions and the task of providing the detailed message flows becomes more difficult as this should be performed separately for each service. For deriving the message flow for IN services, one should use the Global Service Logic (GSL) of the services which is described using the Global Functional Model (GFM). The GFM constitutes a standardized method for representing IN services in terms of Service Independent Blocks (SIBs). The interaction diagram of IN services can be derived from the Global Service Logic translating the chain of SIBs to transactions among functional entities and signaling protocols (see also Chapter 1.2).

The above method, extends the signaling scenarios to include also IN messages. For example a signaling message $x$ will generate an IN message $y$ from B-SSP to B-SCP. Message $y$ will result in the generation of another IN message $w$ which will be forwarded to the B-SSP. Upon reception of message $w$ the B-SSP will generate a signaling message $z$ to the next node of the network. The IN signaling scenario is the sequence of messages $x, y, w, z$. All messages experience a random delay in the specific point of generation and processing. With this approach traffic scenarios can be defined for each component under study, while Poisson arrivals only for call requests.

An alternative way to describe the signaling traffic is to assume Poisson arrivals to each physical or even functional entity. The latter case is widely used in analytical studies due to the well known properties of the Poisson distribution [8, 9].

## 4.2.2   Models for the Intelligent Network Functional Entities and Signaling Protocols

Following the common practice of tele-traffic engineering physical entities are modeled by queuing systems. Depending on the item under study, a physical entity can be modeled as a single queue or a network of queues.

In the single queue approach, the particular type of each message is modeled by virtue of the time required for the processor of the physical entity to handle the message. This time increases with the number of information elements a message contains. Each message is allocated a weight which is the number of its information elements normalized to the minimum number of information elements a message may have. This weight is then multiplied by a Time Unit, which is the time required by the specific processor to perform a 'basic' operation. The Time Unit may vary with the physical entity to model different implementations of user side terminals and network side equipment. The single queue approach is used in the performance evaluation study of Chapter 4.4. A detailed list of the weight of each message involved in a VoD call can be found in Section D.1 of Appendix D.

In the network of queues approach, a more detailed modeling of the protocol operations per message is attempted. A message entering a protocol passes through a number of queues each one representing a function executed by the protocol. A protocol message may trigger the generation of another message destined to a functional entity of the same node or to the peer protocol of an adjacent node. The itinerary a message follows defines the signaling/IN processing scenario. Processing times are now defined per function and the weight of each message becomes equal to the number of intermediate queues in its itinerary. The network of queues approach is used in the performance evaluation studies appearing in Sections 4.3.1 and 4.5.1. Values for the processing times of each function can be found in Section D.1 of Appendix D.

Using the decomposition and aggregation principle [9] functions inside a functional entity or signaling protocol can be studied separately and in this respect the physical entities can be described with accuracy but also simplicity. When analytical methods are applied the system statistics are easily derived using a bottom up approach, that is, performance metrics are calculated for each individual queue and their aggregation gives the corresponding statistic values of the higher level (either the protocol and functional entity or the physical entity). In the following some indicative examples are given for the application of the network of queues approach to the modeling of the signaling protocols and functional entities involved in an IN session.

## 4.2.2.1 Detailed Models

The decomposition principle applied to model in detail the physical entities reduces the model complexity and can be used independently of whether results are obtained by simulation or analysis. On account of the fact that the input traffic for each layer is the output traffic of the adjacent ones, a simulation tool can be efficiently split in separate routines each one implementing a particular protocol layer. Each routine uses as input the output of the routines corresponding to the upper and lower layers. When analytical methods are employed each protocol layer is treated as a separate model and is examined separately. To do so, Poisson arrivals are assumed for each model. A comparison between the above two methods is useful to verify the accuracy of the results obtained by analysis.

The queuing models presented below describe the layer functions as well as the information flow among processing phases of the protocol server. Input and output arcs are used to represent the communication among protocol layers residing in the same or different physical entities. These transitions depend on the information flow derived from the Global Service Logic (GSL) description of a specific service. The GSL is used to describe the specific processes executed for the provision of a particular service. One (or more) processor can serve the entire protocol. Priorities can be assigned to processing phases in order to schedule the process execution and a processing time is assigned for servicing each queue. An example of priority rules applied to different processing phases inside a protocol or functional entity is given in Section D.3 of Appendix D. Processing times are defined according to the message length and the complexity of the executed processes. Information can be exchanged among functions (queues). A message departing from one function can be fed back to another or even to the same one. Analytical solutions for the above system can be obtained by virtue of the M/G/1 queuing system with priorities and feedback. Its analysis is presented in Appendix C.

In the following a set of models is presented for the most typical signaling protocols and functional entities of an Intelligent Broadband Network. These include models for Q.2931 and B-ISUP signaling protocols (see Chapter 1.3) and for CCF, SSF, SRF, SCF and SDF functional entities (see Chapter 2.2). The following assumptions are made: Since the focus of this section is on IN functionality and performance, the transfer network protocols are not modeled in detail. The same holds also for the maintenance, compatibility and management functions of B-ISUP and SRF.

*Q.2931 model.* The Q.2931 model is illustrated in Figure 4.2.1. The call/connection processing control has been divided in two processes, the incoming and outgoing call/connection process.

*B-ISUP model.* The B-ISDN User Part consists of the following Application Service Elements (ASEs): Bearer Connection Control (BCC), Maintenance Control (MC), Call Control (CC), Unrecognized Information (UI) and Single Association Control Function (SACF) and the Application Process (AP). The AP contains call control, maintenance and compatibility functions which are not included in the model. The model of Figure 4.2.1 consists of the BCC, CC ASEs, the SACF and the call control AP. Each one of these have been modeled as a separate processor phase. When the SACF receives a message, it distributes information to ASEs according to the protocol rules. The output from the ASEs is received by the SACF which forwards an appropriate message to the application process or to the underlying MTP-3 protocol layer.

*CCF/SSF model.* The CCF/SSF model of Figure 4.2.2 consists of the Basic Call Manager (BCM), the Feature Interactions Manager/Call Manager (FIM/CM), the Non-IN Feature Manager (NIFM), and the IN-Switching Manager (INSM).

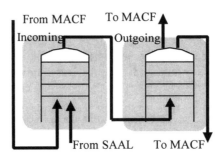

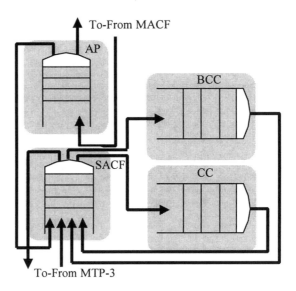

**Figure 4.2.1**
Q.2931 and B-ISUP model

The BCM detects basic call and connection control events that can lead to the invocation of IN service logic instances or should be reported to active IN service logic instances. The INSM interacts with the SCF in the course of providing IN service features to users. It detects IN call/connection processing events that should be reported to active IN service logic instances. The FIM-CM provides mechanisms to support multiple concurrent instances of IN service logic instances and non-IN service logic instances on a single call. The NIFM executes the non-IN service logic instance.

When a user generates a call request it interacts with the CCF/SSF to request the set-up of a call. The BCM creates a Basic Call State Model (BCSM) to represent the basic call control functions required to establish and maintain this call for the user. The BCM sends information to the FIM/CM reporting a BCSM event and the current state of the BCSM in which the event is detected. The FIM/CM receives and processes the BCM event indication to determine if the event is to be processed by an IN service logic instance or a non-IN service logic instance. It also determines if the event should be processed by a new service logic instance or an existing active instance.

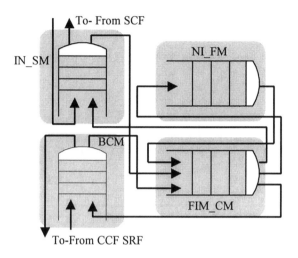

**Figure 4.2.2**
CCF/SSF model

If the event should be processed by an IN service logic instance the FIM/CM sends an IN event indication to the IN-SM that reports a call processing event, the current state of the call in which the event is detected and whether the event should be handled by a new instance of IN service logic or an existing active instance. The IN-SM receives and processes the IN event indication and if a new instance of an IN service logic instance is to be invoked, it creates a new instance of an IN-Switching State Model (IN-SSM) to represent the state of the user's call and connection in a matter accessible to the SCF. If the event is for an existing instance of an IN service logic it updates the state of the existing IN-SSM to reflect the state of the user's connection. It then sends an SSF information flow to the SCF providing a view of the current state of the IN-SSM.

The IN-SM receives an SCF information flow and processes it to manipulate the state of the IN-SSM as requested. It generates an IN control request to the FIM/CM. The FIM/CM receives and processes the IN control request, and sends a BCM control request to the BCM. The BCM receives and processes the BCM control request and manipulates one or more BCSMs to satisfy the request.

*SRF model.* The SRF model of Figure 4.2.3 consists of two components: the Functional Entity Access Manager (FEAM) which provides the necessary functionality to exchange information with other functional entities and the SRF Resource Manager (SRF-RM) which manages resources contained in the SRF. When the FEAM receives a resource request indication, it forwards this to the SRF-RM. This, with the help of the Resources Access Entity (RAE) retrieves the appropriate resources and sends them to the called functional entity or even to the user via the CCF/SSF. The SRF exchanges information with the Service Management Function but since we were not interested in managementfunctions this entity is not included in our model.

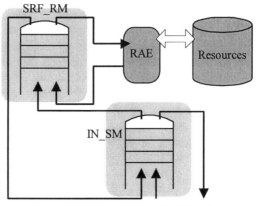

**Figure 4.2.3**
SRF model

*SCF model.* The SCF model of Figure 4.2.4 consists of five sub-models and two libraries. These are the Functional Entity Access Manager (FEAM), the Service Logic Program Manager (SLPM), the Service Logic Execution Manager (SLEM), the Functional Routine Manager (FRM), the SCF Data Access Manager (SCF-DAM) and the Service Logic Program (SLP) and Functional Routine (FR) Libraries.

The FEAM provides the necessary functionality to exchange information with other functional entities and interacts with all the other service managers inside the SCF model. The SLPM manages the reception and distribution function of Service Logic Programs (SLPs) from other entities. The FRM is used for reception and distribution of functional routines to functional routine library via the Library Access Entity (LAE). The SCF-DAM provides the functionality to storage, management and access of shared and persistent information in the SCF and the functionality to access remote information in SDFs. It interacts with the SLEM to provide this functionality. The SLEM handles and controls the total service logic execution and interacts with SCF-DAM. It also has access to SLP and FR libraries via the corresponding access entities in order to support the service logic execution.

When SCF receives an SSF information flow, the FEAM forwards it to SLEM. If a new instance of IN service logic is to be invoked the SLEM selects the SLP for execution of the SLP library. Then it invokes an SLP Instance (SLPI) that realizes the desired service feature. When the SLPI is executed, SLEM invokes functional routines from the FR. In case the execution of the SLPI requires the invocation of specialized resources which are located in remote

SRFs, the SLEM interacts with the SCF-DAM which provides a means to address the appropriate functional entity. When the SLEM processes the SSF information, SCF interacts with the SSF to requests the IN-FM to manipulate the state of the IN-SSM.

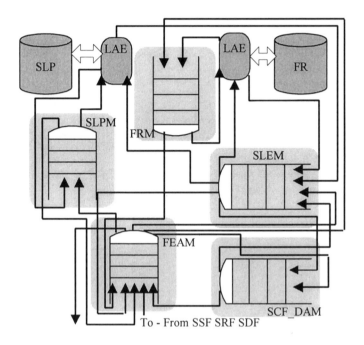

**Figure 4.2.4**
SCF model

*SDF model.* The SDF model consists of two components, the Functional Entity Access Manager (FEAM) which provides the necessary functionality to exchange information with other functional entities and the SDF Data Manager (SDF-DM) which provides the necessary functionality to manage data contained in the SDF. When the FEAM receives a request indication, it forwards this to SDF-DM. This with the help of the data access entity retrieves the appropriate data and returns the appropriate response/result to the called functional entity. It is modeled analogously to the SRF model (Figure 4.2.3).

# References

[1]  Kosal H, Skoog R, *A Control Mechanism to Prevent Correlated Message Arrivals from Degrading Signaling No. 7 Network Performance*, IEEE J. Select. Areas Commun., **12**, 439-445, 1994

[2]  Manfield D, Millsteed G, Zuckerman M, *Performance Analysis of SS7 Congestion Controls Under Sustained Overload*, IEEE J. Select. Areas Commun., **12**, 405-414, 1994

[3]  Duffy D, McIntosh A, Rosenstein M, Willinger W, *Statistical Analysis of CCSN/SS7 Traffic Data from Working CCS Subnetworks*, IEEE J. Select. Areas Commun., **12**, 544-551, 1994

[4]  Zepf J, Rufa G, *Congestion and Flow Control in Signaling System No. 7 - Impacts of Intelligent Networks and New Services*, IEEE J. Select. Areas Commun., **12**, 501-509, 1994

[5]  Smith D, *Effects of Feedback Delay on the Performance of the Transfer-Controlled Procedure in Controlling CCS Network Overloads*, IEEE J. Select. Areas Commun., **12**, 424-432, 1994

[6]  Bolotin V, *Modeling Call Holding Time Distributions for CCS Network Design and Performance Analysis*, IEEE J. Select. Areas Commun., **12**, 433-438, 1994

[7]  Rumsewicz M, *On the Efficacy of Using the Transfer-Controlled Procedure During Periods of STP Processor Overload in SS7 Networks*, IEEE J. Select. Areas Commun., **12**, 415-423, 1994

[8]  Bafutto M, Kuhn P, Willmann G, *Capacity and Performance Analysis of Signaling Networks in Multivendor Environments*, IEEE J. Select. Areas Commun., **12**, 490-500, 1994

[9]  Willmann G, Kuhn P, *Performance Modeling of Signaling System No.7*, IEEE Commun. Mag., **28** , 44-56, 1990

# Chapter 4.3

# PERFORMANCE ANALYSIS OF ALTERNATIVE ARCHITECTURAL SOLUTIONS

The focus of this chapter is on the analysis of two main aspects for the functionality mapping in an Intelligent Broadband Network. The first evaluation aims at analyzing the entire system behavior by considering two alternatives for the mapping of the functional entities of the IN Distributed Functional Plane into physical entities. The purpose is to identify the system bottlenecks and to highlight this architectural alternative that allows the most efficient operation of the system.

The second performance issue faced is relevant to the mapping of the IN functionality that allows the realization of the 'session control domain' (see also Chapter 2.2) into alternative physical settings. In an Intelligent Broadband Network the session control level is realized by a novel IN Switching State Model (IN-SSM) that allows the association and the coordination of the network resources involved in the realization of the IN service. In this chapter two main architectural solutions to realize the above-mentioned coordination by means of the session concept are addressed, and the impact on the performance behavior of the system is evaluated.

Moreover, the impact of the utilization of two different capability sets of the B-ISDN (CS-1[1] and CS-2.1[2]) on the performance of the system, is also considered. The aim of the analysis is to show how possible enhancements in the B-ISDN control and transport capabilities (in particular the handling of point-to-multipoint calls) affect the IN control functions that realize the service. This aspect plays a significant role in the functionality mapping issue. It can be seen as a comparison between scenarios where the IN realizes the whole coordination and association of the calls that supports the service, and scenarios, where some control functionality is transferred into the B-ISDN environment.

## 4.3.1 Mapping of Functional Entities into Physical Entities

To evaluate the performance of alternatives for the mapping of the IN functional entities into the physical ones the Video on Demand (VoD) service as provided in an Intelligent Broadband environment (see Chapter 2.4 for the service description) is analyzed. The physical entities involved in the service provisioning are the user terminal (e.g. a set top box or an ATM PC), the B-Service Switching Point (B-SSP), the B-Service Control Point (B-SCP) and the Video Server (VS) that, in the case of an interactive VoD service, is the repository of the encoded videos that are streamed to the set top box. The IN functional entities that participate in the service provisioning of the VoD are the couple Call Control Function (CCF) / Service Switching Function (SSF), the Service Control Function (SCF), the Specialized Resource Function (SRF) and the Service Data Function (SDF).

As far as the functional mapping issue is concerned, while the CCF and SSF are always located in the B-SSP and the SCF in the B-SCP, the SRF can be found either in an integrated B-SSP or in a separate Broadband-Intelligent Peripheral (B-IP). Similarly the SDF can be located either in the B-SCP or in a separate Broadband-Service Data Point (B-SDP).

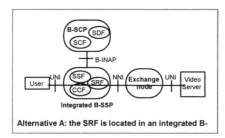

Alternative A: the SRF is located in an integrated B-

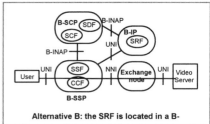

Alternative B: the SRF is located in a B-

**Figure 4.3.1**
Alternatives for the SRF mapping into an Intelligent Broadband Network

In this chapter the emphasis is given to the Specialized Resource Function entity location; this entity has a key role in the VoD service since it allows the user navigation to select the Movie and the Video Server. The emerging alternatives for the generic network configuration with respect to the location of the SRF functional entity are illustrated in Figure 4.3.1.

In alternative *A* the path between the user terminal and the VS involves an integrated B-SSP (where the SRF is mapped), a B-SCP and an Exchange Node. In alternative *B*, the B-SSP is not integrated and the SRF is contained in an ad hoc B-IP.

Apart from the IN functional entities, the physical entities should also contain appropriate signaling protocols enabling the establishment and the release of the bearer connections that support the service.

## 4.3.1.1 Performance Results

The protocols and the functional entities involved in the processing of a call request for a VoD service, are modeled according to the detailed models defined in Subsection 4.2.2.1. The analysis of each model is performed using the M/G/1 queuing system with feedback and non-preemptive service priorities (Appendix C). The traffic parameters used in the analysis are reported in Appendix D. The Rule of Total and External Load described in Section D.3 of Appendix D is used to schedule service priorities among the processes of a single protocol or functional entity. A traffic mix made up of 85% of telephony and 15% of VoD calls is considered.

The performance measure used to capture the behavior of the signaling system is the mean delay experienced by the signaling messages (including queuing and processing) as a function of the calls per second handled by each protocol layer or functional entity during service provisioning. Both alternatives regarding the location of the SRF are considered.

In Figure 4.3.2 the mean delay due to Q.2931 protocol in B-SSP, B-IP and Exchange Node is shown. As it can be noted, in the Exchange Node, the Q.2931 delay is the same for both alternatives, while in the B-SSP the delay is slightly higher for alternative *B*. The reason is that the Q.2931 protocol is additionally activated to cover the communication needs between the B-SSP and the B-IP.

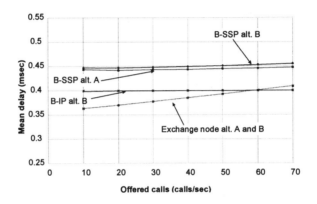

**Figure 4.3.2**
Q.2931 mean delay for B-SSP, B-IP, and Exchange Node

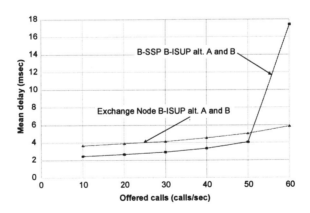

**Figure 4.3.3**
B-ISUP mean delay for B-SSP and Exchange Node

As regards the impact of the two alternatives to the B-ISUP protocol delay in both B-SSP and Exchange Node (Figure 4.3.3) results appear quite stable; this is because the number of messages handled by B-ISUP does not vary with the alternatives.

Figure 4.3.4 presents the mean delay for the IN functional entities (SCF, CCF/SSF and SRF). The SCF mean delay is the same in both alternatives. For the CCF/SSF, a slight difference between the curves is observed which becomes more viable for higher values of offered calls. This difference becomes more evident from a simulative study where the effects of the more bursty arrivals of alternative A can be taken into account [3].

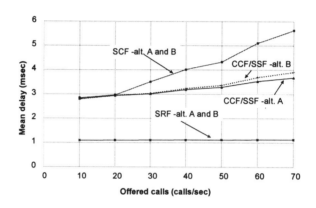

**Figure 4.3.4**
SCF, CCF/SSF, SRF mean delay

As regards the SRF, since the load of the functional entity does not change, the mean delay for both alternatives does not present appreciable differences. The results of Figure 4.3.4 show that the SRF can serve more than 70 calls/s with an acceptable delay per call. On the other hand, as seen in Figure 4.3.3, it is the B-ISUP protocol that determines the maximum number of calls that can be processed by the network (the B-ISUP protocol of the B-SSP restricts this number to 50 calls/s).

The main conclusion that can be drawn from these figures is that there is no significant difference in terms of call set up delay with respect to the location of the SRF. Also the same SRF can be employed to serve more than one B-SSP with acceptable performance. This suggests the utilization of a multi-purpose separate B-IP instead of a complex integrated B-SSP.

## 4.3.2  Alternative Architectural Solutions for the Mapping of the Session Control Functionality

To show the impact of the session concept on the performance behavior of an Intelligent Broadband system, alternative architectural solutions for realizing the calls/connections coordination at the session control level are evaluated. The service considered is the Broadband Video Conference (B-VC) (see also Chapter

2.4). This multimedia multi-point service fully exploits all the advantages brought about by the session concept, since it requires complex control functionality to realize the coordination of all the calls/connections needed for the service delivery.

Figure 4.3.5 shows the considered general network scenarios. In both alternatives the B-SSP, the B-SCP, the B-IP and the Exchange node are present. The B-SSP contains the couple CCF / SSF, the B-IP the SRF, and the B-SCP the SCF and the SDF.

In case the service is entirely realized by a single interaction between the B-SCP and a B-SSP, the corresponding network configuration will comprise all the physical entities appearing in Figure 4.3.5 with the exception of those depicted in gray.

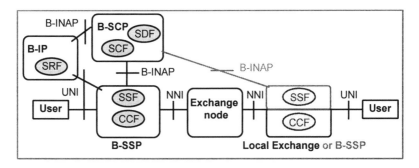

**Figure 4.3.5**
Network configuration model for the mapping of the session control functionality

A more general scenario foresees also the parts shown in gray in Figure 4.3.5 and points out the possibility to have an interaction of the B-SCP, in the context of the same service instance, with two B-SSPs (or more). This multiple interaction has the consequence that the Service Logic interacts with a multiplicity of IN-SSMs. In other words, the IN functionality needed to realize the coordination in the session control domain is mapped into different B-SSPs. Obviously this implies that more then one node in the network supports B-SSP capabilities (in Figure 4.3.5 and in the following analysis this capability is assigned to the local exchange nodes).

The opportunity offered by the mapping of the session concept into different B-SSPs can be advantageous from both an IN and a B-ISDN point of view [4]. A typical case is when the users involved in the service are spread in a wide geographical area and thus they can be very far away from the B-SSP where the service has been started (this is the case analyzed in the following). In this case the approach with only one session can lead to a very inefficient use of the network resources from both a transport and a signaling point of view since all connections are bound to cross a specific network node. In other words the case with only one B-SSP brings to a centralization of all the IN control functionality in a single node that could become a bottleneck of the system, especially when the service requires a large amount of signaling messages to be processed at the B-SSP level.

To overcome this impairment the approach based on a multiple interaction can be used. This approach is named in the following *distributed approach* to indicate the distribution among multiple B-SSPs of the functionality that realize the session control domain. On the other hand the approach based on a single interaction is named *centralized approach.*

## 4.3.2.1 Alternative Architectural Solutions for B-VC Service Provision

To carry out the performance evaluation of the B-VC, the *Conference Establishment* procedure is considered. This procedure takes place after the conference has been created and all conferees have been invited to participate (for more details see Chapter 2.4). The approaches are applied for both the B-ISDN Signaling CS-1 and CS-2.1.

The main steps to realize the conference establishment in the centralized approach are described in the following:

- The first conferee (named conference coordinator) invokes the establishment procedure;
- The SCF sends to the B-SSP subsequent 'SCP-initiated call' commands to realize a fully meshed interconnection of all the invited conferees;
- The SSF modifies the session view, orders the CCF to set-up the desired bearer connections (bi-directional point-to-point or unidirectional point-to-multipoint) and reports the results to the SCF.

Figure 4.3.6 illustrates the network configuration arising from the application of the centralized approach with the use of point-to-point connections (indicated in the following as CPP) in the case of a conference composed by three conferees. When point-to-multipoint connections are used, the centralized approach is indicated as CPMP.

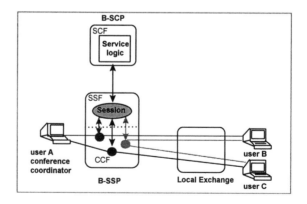

**Figure 4.3.6**
Network configuration for a centralized approach

The bearer connections between each user and the B-IP are not shown; for the sake of neatness only one type of bearer connection (for example the video one) is represented. As depicted in this figure, all the resultant bearer connections are bounded to cross a fixed B-SSP where they are coordinated in a single IN session instance (the arrows in the CCF/SSF represent the interactions between each independent Basic Call State Model and the IN-Switching Manager).

In the case of a distributed architectural solution the service is handled in a multiplicity of sessions; each session holds only a partial view of the service configuration while the overall representation is realized inside the SCF. For the sake of simplicity, in the analysis it has been considered that each local exchange has B-SSP capabilities and that the B-SCP knows the position of the users with respect to the B-SSPs. While the first hypothesis is respected in most of the networks, the second one is more complex to be realized. In practice the second hypothesis can be relaxed, maintaining the validity of the approach, by requiring that the B-SCP has only the capability to identify the location of a user with respect to a geographic area where a reference B-SSP is selected. This information can be acquired by simply looking at the user number prefixes.

The main steps to realize the conference establishment in case of a distributed approach are the following:

- The conference coordinator invokes the establishment procedure in its local B-SSP;
- The SCF selects the B-SSPs where the service should be realized;
- The SCF commands the creation of the new sessions, in addition to that already present in the B-SSP where the conference coordinator has required the execution of the service;
- The SCF sends to the B-SSPs subsequent 'SCP-initiated call' commands to realize a fully meshed interconnection of all the invited users;
- Each SSF modifies its session view, orders the relevant CCF to set-up the required bearer connections (bi-directional point-to-point or unidirectional point-to-multipoint) and reports the results to the SCF;
- The SCF merges the reported results in order to obtain the overall service configuration view.

It is to be noted that, in general, only two main architectural innovations must be introduced to realize a distributed scenario. The former is the possibility to activate a session in an SSF on behalf of the SCF. The latter is a generalization of the representation model of the service in the SCF, designed to support the Service Logic Instance that allows the handling of all the partial views represented by the single sessions.

Figure 4.3.7 represents the architectural scenario in case of a distributed approach. It can be noted that a bearer connection (Fig 4.3.7, users B and C) is added to the new session created in the B-SSP2.

In the following the distributed approach is indicated as DPP or DPMP to distinguish between the use of point-to-point connections and a point-to-multipoint connection.

The introduction of this new approach is justified, from a performance point of view, since the session distribution can be very useful to achieve, as it will be shown in the following, a network resource saving, especially when the users are spread in a wide geographic area; moreover this approach allows to reduce the mean load on each B-SSP. These results can be generalized since the usefulness of the distributed approach is not limited to the B-VC but comprises a large class of services characterized by similar requirements (several parties spread in a network) that would experiment the same resource saving: TV-distribution, Tele-Education, Computer Supported Co-operative Working, etc.

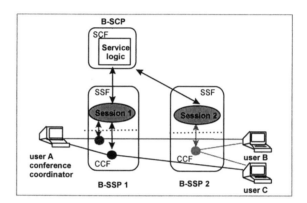

**Figure 4.3.7**
Network configuration in a distributed approach

To compare the approaches presented two main aspects are analyzed: the transport resource utilization and the signaling load on the network nodes. The performance analysis allows us to compare the two architectural solutions (distributed vs. centralized) and to evaluate the impact of the utilization of a B-ISDN point-to-multipoint bearer capability with respect to a point-to-point one.

Several network configurations with a variable number of B-SSPs and conferees (indicated in the following by the parameter $N$) are considered. For a given network configuration all the possible associations between users and B-SSPs are analyzed and numerical results are obtained.

## 4.3.2.2 Performance Results on the Transport Resource Utilization

As far as the transport resources are concerned the cost of the bandwidth allocation is assumed to be the more representative performance parameter. This value is considered linearly dependent on the mutual B-SSPs distance.

The main results are derived by fixing a schematic reference network configuration, depicted in Figure 4.3.8. Six B-SSPs are located at the vertices of two triangles in such a way that the physical distance from a reference point in the center of the configuration is R1 for the first three B-SSPs and R2 for the remaining ones. The actual links between the B-SSPs are not shown.

The rationale behind this choice is to describe a configuration corresponding to a first cluster of B-SSPs relatively close to each other, surrounded by additional B-SSPs located at a greater distance. By varying the above introduced parameters R1 and R2, pros and cons of the centralized approach versus the distributed one can be singled out.

The general expressions used to derive the Cost Value (CV) are the following:

$$CV_{CPP} = 2 \cdot N(1) + \sum_{j=2}^{N_{SSP}} NU(j) \cdot W(1, \ j) + 2 \cdot \sum_{j=2}^{N_{SSP}} \binom{NU(j)}{2} \cdot 2 \cdot W(1, \ j) +$$

$$+ 2 \cdot \sum_{i=2}^{N_{SSP}-1} \sum_{j=i+1}^{N_{SSP}} NU(i) \cdot NU(j) \cdot [W(1, i) + W(1, j)]$$

$$CV_{DPP} = 2 \cdot \sum_{i=1}^{N_{SSP}-1} \sum_{j=i+1}^{N_{SSP}} NU(i) \cdot NU(j) \cdot W(i, j)$$

$$CV_{CPMP} = NU(1) \cdot \sum_{j=2}^{N_{SSP}} W(1, j) +$$

$$+ \sum_{i=2}^{N_{SSP}} NU(i) \cdot \left\{ W(1, i) + \sum_{j=2, j \neq i}^{N_{SSP}} W(1, j) + \delta[NU(i) > 1] \cdot W(1, i) \right\}$$

$$\text{where } \delta[NU(i) > 1] = \begin{cases} 1 & if \quad NU(i) > 1 \\ 0 & if \quad NU(i) = 1 \end{cases}$$

$$CV_{DPMP} = \sum_{i=1}^{N_{SSP}} \sum_{j=1, j \neq i}^{N_{SSP}} NU(i) \cdot W(i, j)$$

Where:
- *NU(i)*, represents the active number of conferees located to the i-th local exchange (in the centralized approach) or located to the i-th B-SSP (in the distributed approach);
- $N_{SSP}$, represents the number of local exchanges (in the centralized approach) or the B-SSPs (in the distributed approach);
- *W(i, j)*, represents the weight assigned to the distance between the i-th and the j-th local exchange (in the centralized approach) or between the i-th and the j-th B-SSP (in the distributed approach).

The measure used to capture the performance behavior of the system (i.e. the cost value) could be assigned by considering also more complex weight assignment criteria.

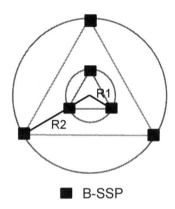

■ B-SSP

**Figure 4.3.8**
Scheme of a generic network configuration

Figures 4.3.9 and 4.3.10 show the results obtained from the analysis of the network configuration depicted in Figure 4.3.8, with R2/R1=100. In Figure 4.3.9, the B-SSP relevant to the conference coordinator is located on a vertex of the internal triangle, while, in Figure 4.3.10, it corresponds to a vertex of the external triangle.

As it can be noted, in the case of point-to-point connections the distributed approach is always very cost-effective with respect to a centralized approach, whereas if point-to-multipoint connections are used, the distributed approach is convenient only if the B-SSP relevant to the conference coordinator is far away from the 'baricentric' position of the network configuration. Finally, the greater is the number of users, the more convenient is the use of point-to-multipoint connections.

These results can be justified by considering that the distributed approach allows users located near the same B-SSP to be interconnected through the B-SSP itself. Moreover, in the case of point-to-multipoint connections, when a user relevant to a B-SSP has to be added to an active conference and a leg versus that B-SSP already exists, no further bandwidth has to be allocated. This is because in this case the cell replication function can be used.

The main general conclusion that can be drawn is that the cost value depends on the position of the B-SSP to which the conference coordinator belongs (named B-SSP$_{coordinator}$). The centralized approach gives acceptable results only if the B-SSP$_{coordinator}$ is located in a 'baricentric' position. In this case the utilization of point-to-multipoint connections, together with the centralized approach, achieves the best resource utilization. On the other hand when the B-SSP$_{coordinator}$ is far from the others the distributed approach is more convenient.

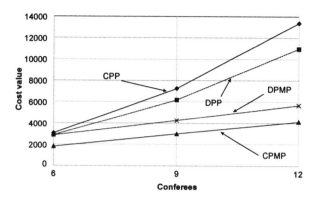

**Figure 4.3.9**
Cost value when the B-SSP relevant to the conference coordinator is located on the vertex of the internal triangle

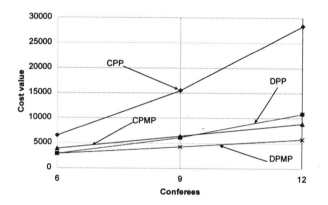

**Figure 4.3.10**
Cost value when the B-SSP relevant to the conference coordinator is located on the vertex of the external triangle

## 4.3.2.3 Performance Results on the Signaling Load

As far as the signaling load is concerned, both the B-ISDN signaling and the IN interactions are analyzed. The performance evaluation is carried out by calculating the number of signaling messages handled by each network element during the realization of the B-VC control procedures. This parameter is related to the processing load supported by the network elements and represents a first measure of the signaling load at the application level (not considering the overhead of the lower layers of the protocol stacks).

As regard the B-ISDN signaling load, the general result (Figure 4.3.11) that can be drawn from the analysis is that for both the bearer capabilities (point-to-point or point-to-multipoint) the distributed approach is very cost effective.

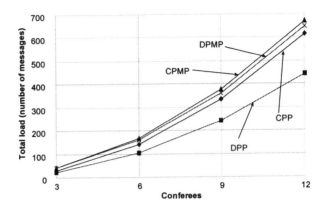

**Figure 4.3.11**
Total mean load per B-SSP due to B-ISDN signaling

Such advantage arises from the saving of signaling messages exchanged at the NNI interfaces (the signaling messages exchanged at the UNI remain the same for both alternatives).

Another result is that the adoption of point-to-point-connections experiences a better performance with respect to the point-to-multipoint ones. This can be justified by considering that, even if the number of connections to be set-up by using point-to-multipoint capabilities is lower than in case of point-to-point ones (2*N vs. N*[N-1] assuming that audio and video will be transferred over separate connections), the signaling messages exchanged for the set-up of a point-to-multipoint connection are proportional to the number of leafs (one 'set up' for the first leaf and N-2 'add party' for the other leafs). As a consequence to set up a single point-to-multipoint connection, that will allow a unidirectional interaction diagram, N-1 signaling messages are sent through the network interfaces.

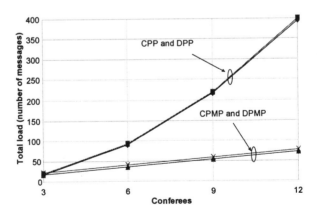

**Figure 4.3.12**
Total number of B-INAP messages exchanged

As far as the total IN signaling load is concerned, a first observation is that the distributed approach does not entail a great increase in the number of messages exchanged by the B-SCP (Figure 4.3.12). A slight difference is determined by the messages required to create new sessions in the B-SSPs and it depends on the number of B-SSPs that are involved in the service instance.

In the case of the distributed approaches, the processing load on the B-SSP$_{coordinator}$ is smaller with respect to the centralized one since some connections are handled in other B-SSPs (Figure 4.3.13).

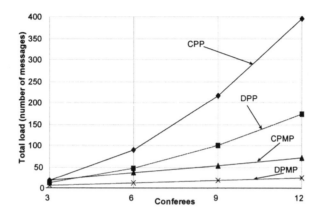

**Figure 4.3.13**
Mean signaling load on B-SSP$_{coordinator}$

The processing load of the B-INAP signaling is smaller when point-to-multipoint connections are used. This result can be explained by considering that the number of the B-INAP messages is proportional to the number of connections that have to be activated (the 'SCP-initiated calls' are 2*N in the case of point-to-multipoint connections and N*[N-1] in the case of point-to-point ones).

## 4.3.3 Conclusions

In this chapter some key architectural and performance issues entailed by the introduction of an Intelligent Broadband platform for the support of multimedia services is investigated. This is accomplished by considering two main aspects related to the functionality mapping issue.

Firstly the performance of the protocols and functional entities involved in the service realization have been captured. Special emphasis is placed on the location of the SRF for which two alternatives are investigated. The results show that for one B-SSP the performance of the system is almost stable independently of whether the SRF is embedded into an integrated B-SSP or located in a standalone B-IP. Since the B-ISUP protocol inside the B-SSP is the one that determines the overall number of supported calls, the B-IP alternative becomes more advantageous since a single standalone B-IP can serve multiple, non integrated (and hence less complex), B-SSPs.

In the second part of the chapter two main architectural solutions for the mapping of the IN functionality that realize the session control domain are analyzed. The first approach, indicated as centralized approach, is based on a single interaction between the B-SCP implementing the service logic and the B-SSP from which the service has been invoked and is relevant to a one to one mapping between the session functionality and the SSF that realizes it. The second foresees a distribution of the session functionality among multiple B-SSPs; this causes the B-SCP to interact with all the involved IN-SSMs during the service execution. Moreover both the CS-1 and the CS-2 step 1 of the B-ISDN signaling are considered. In this context the provisioning of the Broadband Video Conference is evaluated, with particular emphasis on two performance parameters: (i) a cost function representing the transport resources used for the service support; (ii) the signaling load on the network elements.

As regards the performance results on the cost value, in the case of point-to-point connections the distributed approach is always very cost-effective with respect to a centralized approach, whereas if point-to-multipoint connections are used, the distributed approach is convenient only for specific network configurations. As far as the signaling load is concerned, the results highlight that the distributed approach does not entail a significant increase of the total IN signaling load. On the other hand a saving of the B-ISDN signaling is achieved due to a considerable reduction of the signaling information exchanged at the Network Node Interface.

# References

[1]  ITU-T WP 1/11, SG 11, recommendation Q.2931 *B-ISDN User Network Interface Layer 3 specification for Basic Call/Connection Control*, 1994

[2]  ITU-T WP 1/11, SG 11, recommendation Q.2971 *B-ISDN User Network Interface Layer 3 specification for point-to multipoint Call/Connection Control*, 1994

[3]  Kolyvas G, Polykalas S, Venieris I, *A study of specialized resource function mapping alternatives for Integrated IN/B-ISDN architecture*, Proceedings of 2IN'97, Paris, September 1997

[4]  Cuomo F, Listanti M and Pozzi F, *Provision of broadband video conference via IN and B-ISDN integration: architectural and modeling issue*, Proceedings of 2IN'97, Paris, September 1997

# Chapter 4.4

# CONGESTION CONTROL

In this chapter, some congestion control aspects are introduced relevant to the performance evaluation of the Intelligent Broadband Network architecture for the provision of advanced telecommunication services.

For a typical Intelligent Broadband Network architecture, and under the reasonable assumption that the switch will employ very fast processors, the bottleneck of the network is pushed to the B-SCP side. In fact, every IN service instance involves the B-SCP, which has to provide number translation, user authentication and many other service components. In the case of complex services, it also has to configure the service and to coordinate the connections belonging to the same service. This can imply a huge amount of processing and signaling exchange. During particular mass-call periods (televoting, sport events) this entity could be overloaded by a huge amount of call attempts.

If the incoming load at the B-SCP becomes larger than its capacity, as occurs in overload situations, then the performance (e.g. throughput) of the B-SCP degrades and eventually the customers will experience long delays or a high percentage of service refusals (blocked calls).

Some users will abandon their attempts and initiate a reattempt to establish a call. Such reattempts contribute to an increasing load and thus cause the situation to become even worse. Other users will wait until their call request is established. Due to the increased response time due to overload, calls may not reach this state, but may be rejected due to discard or time out.

The objective of the congestion control mechanisms is to tackle these problems. Although congestion may occasionally strike other network elements (B-SSP, B-IP), the study reported here is focused on the B-SCP congestion control study, since this case is by far the more relevant for the IN environment performance assessment.

In the literature, the problem of congestion control has been widely treated with reference to various contexts. Many congestion control mechanisms have been proposed [1-3]. In some papers this problem is discussed with reference to the IN network architecture [4, 5]. In these papers, it is shown that the introduction of an effective congestion control mechanism is mandatory in order to keep an acceptable performance. However, simplifying hypotheses are done as for the offered load and the service modeling. Here, the performance of two congestion control algorithms are studied with reference to the VoD service provision in the Intelligent Broadband Network environment, i.e., accounting for the whole sequence of actions carried out by the B-SCP as the service provision proceeds.

*Intelligent Broadband Networks.* Edited by I. S. Venieris, H. Hussmann
© 1998 John Wiley & Sons Ltd.

In the remainder of this chapter, the two congestion control mechanisms studied will be described; then some results of the simulation study carried out to evaluate the suitability of these mechanisms will be discussed.

## 4.4.1   Description of Congestion Control Mechanisms

In order to compare and analyze B-SCP congestion control mechanisms, it must be first of all defined what 'good' mechanism means. Looking at the objectives of such mechanisms and the context in which they are used, some main criteria can be identified (see also [5, 6]). These are:

- *Robustness*: A B-SCP congestion control mechanism is considered to be robust if it is capable of providing maximum or near maximum call throughput in a variety of overload scenarios. So robustness refers to global 'goodness' of the mechanism in which the sum of the source rates loading with the B-SCP is considered.
- *Fairness*: A B-SCP congestion control mechanism is considered to be fair if it allocates the B-SCP capacity in accordance with some prescribed definition of equity. There are numerous ways to define it more precisely. For example, a particular service could be protected against load bursts caused by other services, or all the services could be assigned the same percentage of computing resources.
- *Reactiveness*: A B-SCP congestion control mechanism should be fast in reacting to varying load conditions.
- *Simplicity*: A B-SCP congestion control mechanism should be easy to implement and administer.

In the frame of the Intelligent Broadband Network, the application of two possible congestion control mechanisms in the developed architecture has been studied. The two selected mechanisms are the *Selective Discard Local Control* (SDLC) and the *Call Gap* (CG). In the following, the two mechanisms are described.

### 4.4.1.1 Selective Discard Local Control (SDLC)

This mechanism operates completely inside the B-SCP, since there is no explicit feedback from the B-SCP to the B-SSP. The B-SCP stops accepting new service requests when it is close to overload conditions. In this way, precious processing resources are preserved for services being already processed, while new services which have no chance to succeed are dropped as they reach the B-SCP.

The service requests discard can be triggered by various conditions; for example, by the number of services currently being treated by the B-SCP, or the number of service requests currently queued in the B-SCP queue, or, more simply, the number of messages currently queued in the B-SCP queue.

The implementation of this mechanism requires a slight overhead in the B-SCP processing, since each incoming message has to be pre-processed before being inserted in the incoming messages queue, in order to single out the new service requests and, eventually, to discard them. As will be shown in the following, this causes some inefficiency from the robustness point of view.

As for the fairness, any discard discipline can be implemented in the B-SCP, allowing to satisfy the chosen fairness criterion; obviously, the more complex the discard discipline, the heavier the implementation. If fields inside the service request message must be accessed in order to check some conditions, further processing is required. Anyway, the basic SDLC mechanism only requires the maintenance of a counter in the B-SCP, and the checking of this counter at each service request arrival.

## 4.4.1.2 Call Gap (CG)

Another basic mechanism for B-SCP congestion control is called Call Gap (CG). Its application to the IN environment has already been studied in the literature [4, 5]. Here, its basic version is described.

There are three variables that determine the mechanism:

$g$: the gap time interval;

$D$: the time interval during which the CG is active;

$N$: the number of service requests that can be accepted during a gap time interval;

The operation of this mechanism involves both the B-SCP and the B-SSP. The idea of CG is to reduce the call attempt rate, i.e. the number of call attempts sent from the B-SSP to the B-SCP per time interval. In other words, the time axis is divided into intervals of duration $g$; at most $N$ call attempts per interval are allowed (the others are discarded).

This reduction only takes place when the mechanism has been activated and lasts for $D$ seconds. In other words, the CG parameters can be chosen to modulate the input stream of incoming call attempts. It allows approximately $(DN / g)$ call attempts within the time interval that the mechanism is active.

Many variants of this mechanism can be implemented; for example, various gapping activation criteria can be applied; moreover, the gap intervals ($g$) can be not adjacent (see for example [3]). In a more sophisticated version, the B-SCP may also control the mechanism by providing information about the level of congestion so that in return the B-SSP can set dynamically appropriate values for $D$, $g$ and $N$.

As example, it is assumed that $g=5$, $D=20$ and $N=1$. On indication of the B-SCP, the CG mechanism is activated by starting the gapping timer ($g$) and the gap duration timer ($D$). When the first call attempt arrives at the B-SSP, it is forwarded to the B-SCP. The call attempts that arrive during the time interval that the gapping timer is less than $g$ will not be passed to the B-SCP. The same check is repeated for all the subsequent gap times, until the gapping duration timer expires (hence equals $D$), then the B-SSP returns to normal operation mode until the B-SCP eventually sends a new message to start the CG mechanism. The mechanism is illustrated in Figure 4.4.1.

It must be considered that, contrarily to the SDLC case, most of the pre-processing required by the CG is carried out in the B-SSP, that has to monitor the income of gapping messages; the B-SCP only has to check some congestion threshold and eventually send a gapping message.

As for the fairness, the same considerations done for the SDLC can be repeated. The implementation of the mechanism is also very simple, since only timers have to be maintained in the B-SSP.

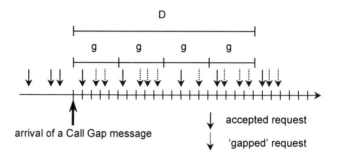

**Figure 4.4.1**
An example of Call Gapping ($g$=5, $D$=20, $N$=1)

## 4.4.2    Performance Evaluation

In the following, some results coming from an extensive simulation campaign on the VoD service are reported. The main objectives of this analysis are to verify the necessity of a congestion control mechanism, to design the various parameters characterizing the analyzed congestion control mechanisms, to compare the analyzed congestion control mechanisms determining pros and cons of each mechanism and to evaluate under which conditions a mechanism is more useful than the other one. In the following, for the sake of brevity, only some results relevant to throughput are discussed.

The model of physical entities adopted for the performance analysis follows the single queue approach described in Section 4.2.2. The users are assumed to offer calls according to a Poisson process; the re-attempts generate a further traffic component. As for the service times, a weight is associated to each message in the manner described in Section 4.2.2. The weight values appear in Section D.1 of Appendix D. The default system parameters adopted for the performance analysis are reported in Table 4.4.1, where 'Mean User Timeout' is the time the user waits before the call is considered unsuccessful in which case the call is repeated. R gives the maximum number of re-attempts.

**Table 4.4.1**
Default system parameters

| Mean User Reaction Time | Mean User Time Out | Number of re-attempts (R) | Propagation Delay | Buffer Size |
|---|---|---|---|---|
| 100 s | 120 s | 5 | 1e-2 s | 1000 messages |

Considering the interaction diagram relevant to the VoD service (see Chapter

2.4), an upper bound for the throughput of the system can be simply determined. In fact, in the overload condition one or more physical entities become a bottleneck for the system. Than the maximum throughput of the congested entity determines the throughput of the whole system. In this study, the Video Server and the set top box cannot be system bottleneck since for them infinite servers are assumed.

The processing time, required in a single physical entity to fulfil a VoD call, can be determined taking in to account the processing time relative to the incoming messages. Since each message has an established weight (see Section D.1 of Appendix D), the sum of the weights associated to these messages multiplied for the Time Unit (amount of time required to perform a 'basic' operation) of the physical entity, provides the processing time spent by the processor of the entity to handle these messages. The pre-processing time has to be added to this time, where needed. So, the above mentioned upper bound is the inverse of the calculated time.

For each physical entity '$x$', Table 4.4.2 shows the number of incoming messages ($N_x$), the total weight ($P_x$) and the consequent maximum throughput

$$M_x = \frac{1}{\left(P_x + N_x \cdot Q_x\right) \cdot T_x}$$ that it can theoretically support, where $T_x$ is its Time

Unit and $Q_x$ is the message pre-processing weight, that is assumed equal to 1 independently of the message type.

The theoretical maximum throughput $Z$ of the system is given by the minimum value of $M_x$; formally,

$$Z = Min_x\left(M_x\right) \qquad (4.4\text{-}1)$$

In order to determine an upper bound function $U_x(L)$ for the generic physical entity $x$, it is supposed that the other entities are not bottleneck. Moreover, no message loss occurs due to buffer overflow. If the offered load $L < M_x$ all the set-up attempts are successful. For $L > M_x$, unsuccessful set-ups could occur; for each of these a processing time equal to $Q_x \cdot T_x$ is spent to pre-process the first message. At each re-attempt the probability of success is equal to $p$. It is assumed that the probabilities of success in two consecutive attempts are independent, due to the high value of the 'User Timeout' appearing in Table 4.4.1. The probability that a call is dropped ($R+1$ successive failures) is given by $(1-p)^{R+1}$. By definition this is also equal to $(L-U_x(L))/L$, i.e.:

$$\left(1-p\right)^{R+1} = \frac{L - U_x}{L}$$

This is an ideal situation where there is not any unsuccessful set-up due to message loss, but a set-up can be discarded only by the congestion control mechanism. The upper bound $U_x(L)$ of the entity $x$ can be evaluated limiting the entity utilization $\rho_x$ to 1. The following equation can be numerically solved with respect to $U_x(L)$:

$$\rho_x = (R+1) \cdot Q_x \cdot T_x \cdot (L - U_x) + \sum_{i=0}^{R} p \cdot (1-p)^i \cdot L \cdot (P_x + (N_x + i) \cdot Q_x) \cdot T_x = 1$$

$$(4.4\text{-}2)$$

under the hypothesis that $(P_x + N_x \cdot Q_x) \cdot T_x > (R+1) \cdot Q_x \cdot T_x$ , while $U_x(L)=0$ for $L \geq M_x$.

The upper bound $B(L)$ of the system is the minimum of the single upper bounds $U_x(L)$, i.e.

$$B(L) = \underset{x}{Min}\big(U_x(L)\big) \tag{4.4-3}$$

**Table 4.4.2**
Static load in each physical entity x

| Physical entity - $x$ | SCP | SSP | IP | STB | VS |
|---|---|---|---|---|---|
| Number of incoming messages ($N_x$) | 14 | 28 | 11 | 11 | 6 |
| Total Weight ($P_x$) | 82 | 103 | 33 | 10 | 18 |
| Time Unit ($T_x$ - ms) | 1 | 0.4 | 1 | 1 | 1 |
| Maximum Throughput ($M_x$ – calls/s) | 12.20 | 24.27 | 30.30 | $\infty$ | $\infty$ |

In the case of uncontrolled system, the theoretical maximum throughput $Z_{UNC}$ is given by Eq.(4.4-1), taking into account that in this case no pre-processing is carried out, and therefore all $Q_x$ are null. Utilizing the value of Table 4.4.2, $M_{UNC}$ = 12.2 VoD call/s. In this case, the upper bound $B_{UNC}(L)$ is independent of $L$ and equals $Z_{UNC}$.

## 4.4.2.2 Performance of the SDLC

The SDLC mechanism is implemented by defining a threshold $S$ inside the B-SCP buffer, so that if more than $S$ overall messages are queued inside the B-SCP, no further service requests are accepted.

For the evaluation of the theoretical maximum throughput in the SDLC case, the effect of the pre-processing inside the B-SCP has to be considered. The theoretical maximum throughput $Z_{SDLC}$ and the upper bound $B_{SDLC}(L)$ are given by Eq. (4.4-1) and Eq. (4.4-3), respectively, with $Q_{SCP}=1$ and all the remaining $Q_x=0$.

Figure 4.4.2 shows the throughput versus the offered load. The upper bounds $B_{UNC}(L)$ and $B_{SDLC}(L)$ and the simulation throughput without congestion control mechanism are compared with the throughputs of the SDLC mechanism, relevant to different values of threshold $S$.

As it can be noted, in the overload region the system throughput provided by the SDLC mechanism is rather independent of the fixed threshold value; for any value of $S$, the throughput curves have the same slope of the upper bound $B_{SDLC}(L)$, and they are quite close to this limit. It can be noted that for offered load around $Z_{UNC}$, the introduction of the SDLC mechanism entails some inefficiency with respect to the uncontrolled system, due to the pre-processing of messages inside the B-SCP. However, the benefits in term of throughput are evident increasing the offered load.

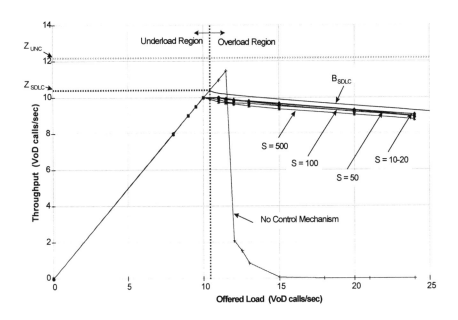

**Figure 4.4.2**
Throughput vs. offered load for SDLC

## 4.4.2.3 Performance of the CG

In this chapter, some results of the simulation study relevant to the Call Gap are reported. A simple version of the CG is implemented, where the gapping is triggered by the excess of a threshold $S$ in the B-SCP buffer. The gapping intervals are adjacent, and $N=1$. Moreover, the B-SCP only communicates the presence of congestion, i.e., no dynamic parameters setting is allowed.

For the Call Gap mechanism, the maximum throughput ($Z_{CG}$) and the upper bound ($B_{CG}(L)$) are evaluated from Eq. (4.4-1) and Eq. (4.4-3), respectively, with $Q_{SSP}=1$ and all the remaining $Q_x=0$. In fact, the pre-processing is done in the B-SSP. Moreover, since the bottleneck is the B-SCP, $Z_{CG} = Z_{UNC}$.

Figure 4.4.3 shows the throughput of the system versus the offered load for different values of $D$ both with the CG mechanism and without any congestion control mechanism. The threshold positioning is fixed to $S = 50$ messages while the transmission rate $RG=1/g$ is fixed to 10% of the maximum throughput ($RG=0.1\ Z_{UNC}$). It can be noted that the maximum throughput is almost reached if a proper setting of $D$ is carried out.

In fact, inefficiency can arise if high values for $D$ are chosen. If $D$ is less than 1.6 s, no throughput degradation is observed. The inefficiency is caused by the fact that, if the gapping period is too long, the B-SCP buffer can become empty, causing an under-utilization of the B-SCP processor. This effect can be also avoided increasing the threshold value, or increasing the transmission rate.

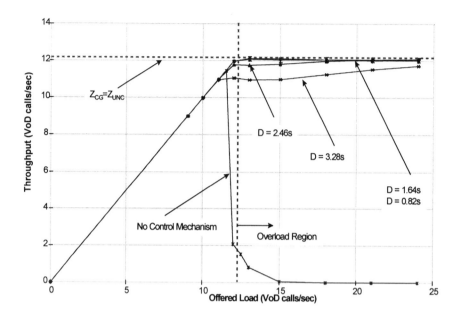

**Figure 4.4.3**
Throughput vs. Offered Load for CG; RG = 0.1 $M_{UNC}$, S = 50

## 4.4.3   Conclusions

Two B-SCP congestion control mechanisms have been described and analyzed by means of a high level simulation model. The first one, called SDLC, foresees the discard of service request messages when the B-SCP is in a congestion state. Though this mechanism allows the throughput in the overload region to be increased, it shows an inefficient behavior due to the amount of messages pre-processing in the B-SCP. This situation is worsened by the user call re-attempts.

The second mechanism is called Call Gap and does not require pre-processing in the B-SCP. The B-SCP has to evaluate its buffer occupancy and to notify to the B-SSP a congestion status when its buffer exceeds a pre-defined threshold. Consequently, the theoretical maximum throughput of the system is actually reached and good performance is obtained. This study was carried out under the hypothesis that the bottleneck of the system is the B-SCP; then, the pre-processing in the B-SSP doesn't cause any problem. In this case, the pre-processing, needed to recognize the Call Gap messages, doesn't influence the throughput because of light load in the B-SSP. If both the B-SCP and the B-SSP become overloaded, the pre-processing entails a reduction of the B-SSP maximum throughput. Therefore, the considered congestion control mechanisms become equivalent.

When the offered load becomes considerable both in the B-SSP and in the B-SCP, a congestion control mechanism obtained by the 'superposition' of the two proposed mechanisms could be the right choice. In particular, in order to protect the B-SCP, the Call Gap mechanism can be used. This means that a pre-processing to recognize the incoming messages is applied in the B-SSP. Then, the application of the SDLC mechanism to protect the B-SSP does not cause any further pre-processing inside this node, since the message discrimination is done once and for all. So, only a further threshold must be implemented in the B-SSP to discard the set-up messages when the B-SSP buffer occupancy exceeds this new threshold.

## References

[1]  Farel R, Gawande M, *Design and analysis of overload control strategies for transaction network databases*, Proceedings of ITC 13, Copenhagen, 1991

[2]  Turner P, Key P, *A new call gapping algorithm for network traffic management*, Proceedings of ITC 13, Copenhagen, 1991

[3]  Berger A, *Comparison of Call Gapping and Percent Blocking for overload control in distributed switching systems and telecommunication networks*, IEEE Trans. Commun., **39**, 574-580, 1991

[4]  Hebuterne G, Romoeuf L, Kung R, *Load regulation schemes for the Intelligent Network*, Proceedings of ISS'90, Stockholm, May 1990

[5]  Smith D, *Ensuring robust call throughput and fairness for SCP overload controls*, IEEE/ACM Trans. Netw., **3**, 538-548, 1995

[6]  Tsolas N, Abdo G, Bottheim R, *Performance and overload considerations when introducing IN into an existing network*, Proceedings of International Zurich Seminar on Digital Communications, Intelligent Networks and their applications, 1992

# Chapter 4.5

# SCALABILITY

Scalability is one of the most important factors in the design of a distributed multimedia system (see also [1-4]). The system must be able to sustain a large number of users and a varying amount of data without any problems regarding the availability of resources and the system performance. In this respect the Intelligent Broadband Network, which is a typical paradigm of such a system due to the distribution of network intelligence in several physical entities, must be properly designed to support the offered services at the required quality. Network scalability can be defined as the ability to increase the 'size' of the network, in some sense, while maintaining Quality of Service and network performance criteria [5]. The 'size' of the network may relate to one of the following:

- *The number of users that must be supported by a network node:* increasing the number of users that must be supported by a certain physical entity can cause serious performance problems because of processing capacity and memory limitations;
- *The number of network nodes and links:* the growth of number of nodes and links may cause an increase on the offered load to a given physical entity, since the physical entity will have to manage the intercommunication with the additional nodes in a more complex topology;
- *The geographical spread covered by the network:* increasing the geographical area that is covered by a network while keeping the number of nodes and links constant will cause an increase of the message propagation delays since these delays are proportional to the length or number of the physical communication links;
- *The number of services provided by the network:* increasing the number of services that a network will have to provide will cause an increase in the offered load and its variability to a given physical entity, since the physical entity will have to support the additional services of different requirements;
- *The size of the data objects:* particularly in some cases like video and audio the size of transmitted files is too large and thus strains the network and I/O capacity causing scalability problems;
- *The amount of accessible data:* the increasing amount of accessible data, makes data search, access, and management more difficult and therefore causes storage and processing problems.

In general performance increases with the number of employed processors [6] while a threshold is identified beyond which any further increase in the number of processors does not enhance the speed of execution. The value of this

*Intelligent Broadband Networks.* Edited by I. S. Venieris, H. Hussmann
© 1998 John Wiley & Sons Ltd.

threshold depends on factors such as the communication between the processors and the load balance.

In this chapter, two network scalability studies are accomplished. The performance of an Intelligent Broadband Network is investigated when the physical entities have dedicated processors for each protocol or functional entity (scenario A) and when a single processor serves the entire physical entity (scenario B). In scenario B the processor speed considered is the one that results in protocol mean delay equal to the mean delay of the multiple processors scenario for low load (i.e. 10 calls/sec). In the second study, the scalability of an Intelligent Broadband Network is investigated when the number of nodes and links increases while the network utilization, i.e., the busy percentage, remains constant.

## 4.5.1    Increasing the Number of Processors

To evaluate the performance of the multiple processor scenario, the detailed models described in Subsection 4.2.2.1 are used. A VoD service is considered and the parameter values are those used in Section 4.3.1 and reported in Section D.1 of Appendix D. For the single processor scenario, the physical entities are modeled as a single queue and the processing time for each message of a protocol or functional entity ($T_{pf}$) is assumed, equal to the mean delay of the corresponding protocol or functional entity as it has been estimated by simulation and analysis studies performed for the network of queues approach for a low load of 10 calls/s (see the corresponding values of $T$ in Tables D.1 to D.6 of Appendix D). The $T_{pf}$ values are given in Table 4.5.1. The queuing system used for both the single and multiple processor scenario is the one appearing in Appendix C. Priorities to protocols and functional entities ($pr$) are assigned according to the Rule of Total and External Load appearing in Section D.3 of Appendix D and used in Section 4.3.1.

**Table 4.5.1**

Protocol and functional entity processing time $T_{pf}$ (ms)/protocol and functional entity priority $pr$

| B-SSP | | B-SCP | | B-IP | | EXC | |
|---|---|---|---|---|---|---|---|
| prot. | $T_{pf}/pr$ | prot. | $T_{pf}/pr$ | prot. | $T_{pf}/pr$ | prot. | $T_{pf}/pr$ |
| Q2931 | 0.45/3 | SCF | 2.7/1 | Q2931 | 0.5/1 | Q2931 | 0.4/1 |
| BISUP | 2.5/1 | SDF | 1.2/2 | SRF | 1.1/2 | BISUP | 3.45/2 |
| SSF/CCF | 2.55/2 | | | | | | |

In analogy to the study presented in Section 4.3.1 a traffic mix made up of 85% of telephony and 15% of VoD is considered and the same performance measure of mean setup delay is used.

In Figure 4.5.1 the VoD mean set-up delay is shown as a function of the number of calls/s. The slopes in the delay curves determine the system bottleneck which as shown in Section 4.3.1 is the B-SSP. Therefore a B-SSP with a single processor (scenario B) can support up to 30 calls/s, while this number increases to 60 calls/s when multiple processors are considered (scenario A).

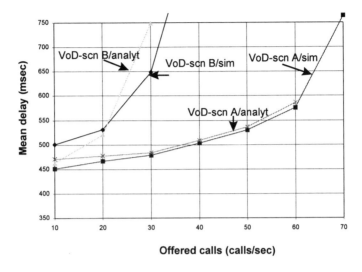

**Figure 4.5.1**
VoD set-up delay

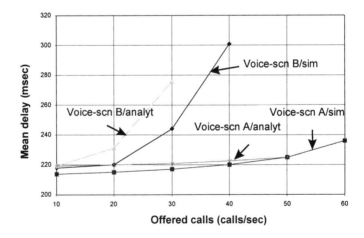

**Figure 4.5.2**
Telephony set-up delay

The utilization of the processors serving the B-SCP and B-IP protocols is low when multiple processors are employed. Therefore to increase the system scalability in terms of the maximum number of calls the system can process one should employ multiple processors for the B-SSP and a single processor for the other network components. It is worth noting to note in both Figures 4.5.1 and 4.5.2 the differences appearing between results obtained by simulation (continuous line) and by analysis. These are due to the assumption of Poisson arrivals adopted by the model of Appendix C, while in reality messages are highly correlated as they linked together to a predefined chain of protocol states and actions. This difference becomes more pronounced in the case of the multiple processor architecture where modeling of the system has been based on the more accurate network of queues approach.

## 4.5.2    Increasing the Number of Nodes

The scalability of an Intelligent Broadband Network is further investigated by increasing the number of nodes and links, while keeping the utilization constant At first more than one interconnected IN islands are assumed (see Figure 4.5.3). Each island consists of four B-SSPs that share a single B-SCP. Two or more B-SSPs belonging to the same or different islands can intercommunicate in the way appearing in Figure 4.5.4.

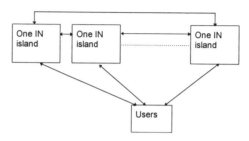

**Figure 4.5.3**
An Intelligent Broadband Network

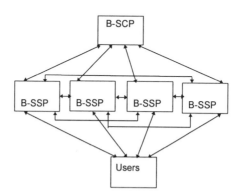

**Figure 4.5.4**
An IN island

Each physical entity (i.e., B-SSP, B-SCP) is modeled by a single server queue with an infinite buffer. The B-VPN service described in Chapter 2.4 is considered, for which processing time measurements have become possible on the experimental prototype network described in Chapter 2.8. All processing times and results are normalized to the value of the Service Request message processing time. This value is used in the remaining parts of this chapter as a Time Unit. The (normalized) processing times appear in Tables D.11 to D.13 of Appendix D.

The following assumptions are made: The load on the system consists of 15% B-VPN calls and 85% telephony. The Time-Based priority scheduling discipline described in Section D.3 of Appendix D is used among all types of requests. In the experiments a 2 Mbit/s transmission bandwidth on the unidirectional links between the B-SSPs, and between each B-SCP and each B-SSP is used. The 2 Mbit/s transmission link is modeled as a single server queue with an infinite buffer and a constant service time (0.00728 Time Units) per message. The total load on the Intelligent Network is assumed to be equally divided among all B-SSPs. The performance measures that are used to capture the performance of the signaling system are the network throughput and the B-VPN call mean setup delay. The B-VPN call mean setup delay is defined as the time duration from the instant when the initiating B-VPN user starts the B-VPN call set-up procedure, until the call set-up procedure is completed, excluding any user response times.

The simulation experiments are accomplished in the following way. First the bottleneck of the network is found by increasing the load (e.g., increasing the number of users per B-SSP). Afterwards this bottleneck is removed by balancing the processing speed of all Intelligent Network physical entities, such that the utilization of each B-SCP and each B-SSP, respectively, are approximately equal. It is found that initially the B-SSPs are the bottlenecks of the network. Therefore the processing speed of each B-SSP is multiplied by the factor 1.4684. The balanced network is then used to accomplish the actual scalability experiments. The experiments were performed in a number of subsequent steps. During each step the number of islands $N$ is increased by a certain factor. Note that initially a fixed number of users is connected to an island. This number is only changed in order to (approximately) adjust the utilization of the network physical entities (e.g., B-SSP and B-SCP) to the chosen value of 0.9.

The initial value of $N$ is one and the maximum value of $N$ was set to 10. For each step, the throughput for a given utilization (i.e., 0.9) and the corresponding B-VPN mean setup delay are estimated.

### 4.5.2.1 Performance Results

This section presents the results obtained for the second set of experiments where the number of IN islands is increased, while keeping the utilization fixed to 0.9. The telephony service is used as background load to reflect a realistic situation and therefore only the obtained results related to the IN (B-VPN) service are discussed. The simulation results are plotted in Figures 4.5.5 and 4.5.6, for the network throughput and B-VPN mean setup delay, respectively. The total network throughput increases (almost) linearly with the number of islands, $N$. The slope of this line is approximately 2.7. The normalized B-VPN mean setup delay (i.e., in Time Units) decreases by increasing the number of islands $N$. The main reason of this is that the input load on each island has to be slightly decreased (in order to adjust the utilization to 0.9) when the number of islands $N$ is increased.

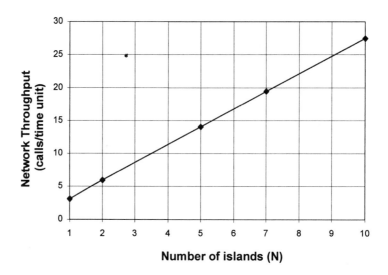

**Figure 4.5.5**
Total network throughput as function of the number of islands

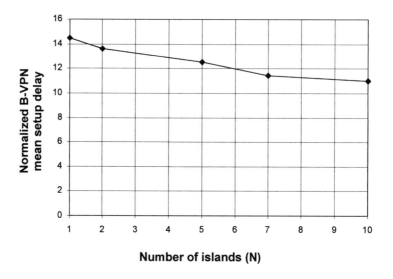

**Figure 4.5.6**
B-VPN normalized mean setup delays as function of the number of islands

# References

[1] Gauthier E., Le Boudec J, *Scalability Enhancements on Connection Oriented Networks*, Lecture Notes in computer science, **1044**, 27-38, 1996

[2] Lin X, Orlowska M, Zhang Y, *Database Placement in Communication Networks for Minimizing the Overall Transmission Cost*, Mathematical and Computer Modelling Magazine, **19**, 7-20, 1994

[3] Martini P, Ottensmeyer J, *On the Scalability of the Demand-Priority LAN - A Performance Comparison to FDDI for Multimedia Scenarios*, Lectures Notes in computer science, **1044**, 146-60, 1996

[4] Saha D, Mukherjee A, *Design of Hierarchical Communication Networks under NodeLink Failure Constraints*, Comp. Commun., **18**, 378-83, 1995

[5] Karagiannis G, van Beijnum B, Niemegeers I, *On the Integration of the UMTS and B-ISDN System*, Proceedings of IFIP (HPN'97), Chapman & Hall, 1997

[6] Ghosal D, Lakshman T, Huang Y, *Parallel Architectures for Processing High Speed Network Signaling Protocols*, IEEE/ACM Trans. Netw., **3**, 1995

# Appendices

# Appendix A

# PROTOCOL STANDARDS

This appendix contains a list of Recommendations about Intelligent Networks and B-ISDN signaling that have been approved or are in drafting phase by ITU-T SG11.

## A.1 Intelligent Network Protocols

In Tables A.1-3 the set of Recommendations about Intelligent Network approved or in drafting phase by ITU-T SG11 are listed.

They are divided in 3 groups:

- Table A1: list of approved Recommendations which provide the foundation for the definition of IN capability sets;
- Table A2: list of approved Recommendations for the IN Capability Set 1;
- Table A3: list of approved Recommendations for the IN Capability Set 2.

**Table A.1**
Recommendations which provide the foundation for the definition of IN capability sets

| Number | Title |
|--------|-------|
| Q.1200 | Q-series Intelligent Network Recommendation Structure |
| Q.1201 | Principles of Intelligent Network Architecture |
| Q.1202 | Intelligent Network Service Plane Architecture |
| Q.1203 | Intelligent Network Global Functional Plane Architecture |
| Q.1204 | Intelligent Network Distributed Functional Plane |
| Q.1205 | Intelligent Network Physical Plane Architecture |
| Q.1208 | General Aspects of the Intelligent Network Application Protocol |
| Q.1209 | Glossary of Terms used in the Definition of Intelligent Networks |

**Table A.2**
Recommendations for the IN Capability Set 1:

| Number | Title |
|--------|-------|
| Q.1210 | Q.121x-series Intelligent Network Recommendation Structure |
| Q.1211 | Introduction to Intelligent Network Capability Set 1 |
| Q.1213 | Global Functional Plane Architecture for Intelligent Network CS 1 |
| Q.1214 | Distributed Functional Plane for Intelligent Network CS 1 |
| Q.1215 | Physical Plane Architecture for Intelligent Network CS 1 |
| Q.1218 | Interface Recommendation for Intelligent Network CS 1 |
| Q.1219 | Intelligent Network Users Guide for CS 1 |

*Intelligent Broadband Networks*. Edited by I. S. Venieris, H. Hussmann
© 1998 John Wiley & Sons Ltd.

**Table A.3**
Recommendations for the IN Capability Set 2:

| Number | Title |
|--------|-------|
| Q.1220 | Q.122x-series Intelligent Network Recommendation Structure |
| Q.1221 | Introduction to Intelligent Network Capability Set 2 |
| Q.1223 | Global Functional Plane Architecture for Intelligent Network CS 2 |
| Q.1224 | Distributed Functional Plane for Intelligent Network CS 2 |
| Q.1225 | Physical Plane Architecture for Intelligent Network CS 2 |
| Q.1228 | Interface Recommendation for Intelligent Network CS 2 |
| Q.1229 | Intelligent Network Users Guide for CS 2 |

## A.2   Signaling Protocols

In Tables A.4-9 the set of Recommendations about B-ISDN signaling approved or in drafting phase by ITU-T SG11 are listed.
They are divided in 6 groups:

- Table A4: list of the Recommendations (all approved) of Capability Set 1 about general aspects and inter-working issues, shared both by access side (DSS2) and network side (B-ISUP);
- Table A5: list of the Recommendations (all approved) of Capability Set 1 for access signaling (DSS2);
- Table A6: list of the Recommendations (all approved) of Capability Set 1 for network signaling (B-ISUP);
- Table A7: list of the Recommendations (all approved) of Capability Set 1 for the layers underlying application protocols, that is Signalling ATM Adaptation Layer (SAAL) and Broadband Message Transfer Part (MTP-3B);
- Table A8: list of the Recommendations (partly approved) of Capability Set 2 for access signaling (DSS2);
- Table A9: list of the Recommendations (partly approved) of Capability Set 2 for network signaling (B-ISUP).

All tables refer to the status of the work of WP1 of ITU-T SG11 after September 1997 plenary.

**Table A.4**
Recommendations defined for Broadband Signaling Capability Set 1: general and common aspects of B-ISDN application protocols for access signaling and network signaling and inter-working

| Number | Title |
|--------|-------|
| Q.2010 | Broadband integrated services digital network overview - Signaling capability set 1, release 1 |
| Q.2610 | Usage of cause and location in B-ISDN user part and DSS2 |
| Q.2650 | Inter-working between Signaling System No. 7 Broadband ISDN User Part (B-ISUP) and digital subscriber Signaling System No. 2 (DSS2) |
| Q.2660 | Inter-working between Signaling System No. 7 - Broadband ISDN User Part (B-ISUP) and Narrow-band ISDN User Part (N-ISUP) |

**Table A.5**

Recommendations defined for Broadband Signaling Capability Set 1: B-ISDN application protocols for access signaling

| Number | Title |
|--------|-------|
| Q.2931 | Digital Subscriber Signaling System No. 2 (DSS2) - User network interface (UNI) layer 3 specification for basic call/connection control |
| Q.2951.1 | DSS2 - Direct Dialing In (DDI) supplementary service |
| Q.2951.2 | DSS2 - Multiple Subscriber Number (MSN) supplementary service |
| Q.2951.3 | DSS2 - Calling Line Identification Presentation (CLIP) supplementary service |
| Q.2951.4 | DSS2 - Calling Line Identification Restriction (CLIR) supplementary service |
| Q.2951.5 | DSS2 - Connected Line Identification Presentation (COLP) supplementary service |
| Q.2951.6 | DSS2 - Connected Line Identification Restriction (COLR) supplementary service |
| Q.2951.8 | DSS2 - Sub-addressing (SUB) supplementary service |
| Q.2955.1 | DSS2 - Closed User Group |
| Q.2957 | DSS2 - User-user signaling supplementary service |

**Table A.6**

Recommendation defined for Broadband Signaling Capability Set 1: B-ISDN application protocols of the network

| Number | Title |
|--------|-------|
| Q.2730 | Signaling System No. 7 B-ISDN user part (B-ISUP) - Supplementary services |
| Q.2735.1 | Signaling System No. 7 B-ISDN user part (B-ISUP) - Section 1: Closed User Group (CUG) |
| Q.2761 | Functional description of the B-ISDN user part (B-ISUP) of Signaling System No. 7 |
| Q.2762 | General functions of messages and signals of the B-ISDN user part (B-ISUP) of Signaling System No. 7 |
| Q.2763 | Signaling System No. 7 B-ISDN user part (B-ISUP) of Signaling System No. 7 - Formats and codes |
| Q.2764 | Signaling System No. 7 B-ISDN user part (B-ISUP) of Signaling System No. 7 - Basic call procedures |

**Table A.7**

Recommendations defined for Broadband Signaling Capability Set 1: Signaling ATM Adaptation Layer (SAAL) and signaling network protocols

| Number | Title |
|--------|-------|
| Q.2100 | B-ISDN signaling ATM adaptation layer (SAAL) overview description |
| Q.2110 | B-ISDN ATM adaptation layer - Service specific connection oriented protocol (SSCOP) |
| Q.2119 | B-ISDN ATM adaptation layer - Convergence function for SSCOP above the frame relay core service |
| Q.2120 | B-ISDN meta-signaling protocol |
| Q.2130 | B-ISDN signaling ATM adaptation layer - Service specific coordination function for support of signaling at the user-network interface (SSCF at UNI) |
| Q.2140 | B-ISDN signaling ATM adaptation layer - Service specific coordination |

| | function for support of signaling at the network node interface (SSCF at NNI) |
|---|---|
| Q.2144 | B-ISDN signaling ATM adaptation layer (SAAL) - Layer management for the SAAL at the network node interface (NNI) |
| Q.2210 | Message transfer part level 3 functions and messages using the services of ITU-T Recommendation Q.2140 |

**Table A.8**
Recommendations defined for Broadband Signaling Capability Set 2: B-ISDN application protocols for access signaling

| Number | Title | Status |
|---|---|---|
| Q.2932.1 | Digital Subscriber Signaling System No. 2 (DSS2) - Generic functional protocol: Core functions | approved |
| Q.2932.2 | Digital Subscriber Signaling System No. 2 (DSS2) - Generic functional protocol: Additional constructs | not approved |
| Q.2933 | Digital Subscriber Signaling System No. 2 (DSS2) - Signaling specification for Frame Relay service | approved |
| Q.2939.1 | Digital Subscriber Signaling System No. 2 (DSS2) - Application of DSS2 service related information elements | approved |
| Q.2941.1 | Digital Subscriber Signaling System No. 2 (DSS2) - User generated identifiers | approved |
| Q.2941.2 | Digital Subscriber Signaling System No. 2 (DSS2) - User generated identifiers (additional) | not approved |
| Q.2959 | Digital Subscriber Signaling System No. 2 (DSS2) - Call priority | approved |
| Q.2961.1 | Digital Subscriber Signaling System No. 2 (DSS2) - Additional traffic parameters | approved |
| Q.2961.2 | Digital Subscriber Signaling System No. 2 (DSS2) Support of ATM Transfer capability in the broadband bearer capability information element | approved |
| Q.2961.3 | Digital Subscriber Signaling System No. 2 (DSS2) - Support of the Available Bit Rate (ABR) ATM Transfer Capability | approved |
| Q.2961.4 | Digital Subscriber Signaling System No. 2 (DSS2) - Support of the ATM Block Transfer (ABT) ATM Transfer Capability | approved |
| Q.2961.5 | Digital Subscriber Signaling System No. 2 (DSS2) - Cell Delay Variation (CDV) tolerance | not approved |
| Q.2961.6 | Digital Subscriber Signaling System No. 2 (DSS2) - Global tagging | not approved |
| Q.2962 | Digital Subscriber Signaling System No. 2 (DSS2) - Connection characteristics negotiation during call/connection establishment phase | approved |
| Q.2963.1 | Digital Subscriber Signaling System No. 2 (DSS2) - Connection modification: Peak cell rate modification by the connection owner | approved |
| Q.2963.2 | Digital Subscriber Signaling System No. 2 (DSS2) - Modification procedures for sustainable cell rate parameters | not approved |
| Q.2963.3 | Digital Subscriber Signaling System No. 2 (DSS2) - | not approved |

|          | Connection modification with negotiation | |
|----------|------------------------------------------|----------|
| Q.2964.1 | Digital Subscriber Signaling System No. 2 (DSS2) - Basic look ahead | approved |
| Q.2971 | Digital Subscriber Signaling System No. 2 (DSS2) - User-network interface layer 3 specification for point to multi-point call/connection control | approved |
| Q.298x | Digital Subscriber Signaling System No. 2 (DSS2) - Support of point to point Multi-connection calls; it is divided into three documents: Q.29cc (Multi-connection Call Control, Q.29bb (Multi-connection Bearer Control) and Q.29pn (Pre-negotiation) | not approved |
| Q.29svp | Digital Subscriber Signaling System No. 2 (DSS2) - Switched virtual path call/connection establishment | not approved |
| Q.293x | Generic concepts for support of multi-point and multi-connection calls | not approved |
| Q.29QoS | Digital Subscriber Signaling System No. 2 (DSS2) - Quality of service | not approved |
| Q.29lij | Digital Subscriber Signaling System No. 2 (DSS2) - Point to Multi-point Leaf initiated join | not approved |

**Table A.9**

Recommendations defined for Broadband Signaling Capability Set 2: B-ISDN application protocols of the network

| Number | Title | Status |
|--------|-------|--------|
| Q.2721.1 | B-ISDN user part - Overview of the B-ISDN Network Node Interface Signaling Capability Set 2, Step 1 | approved |
| Q.2722.1 | B-ISDN user part - Network Node Interface specification for point to multi-point call connection control | approved |
| Q.2722.2 | B-ISDN user part - Network Node Interface specification for multi-connection call/connection control | not approved |
| Q.2723.1 | B-ISDN user part - Support of additional traffic parameters for Sustainable Cell Rate and Quality of Service | approved |
| Q.2723.2 | B-ISDN user part - Support of ATM Transfer Capability (ATC) | approved |
| Q.2723.3 | B-ISDN user part - Support of the Available Bit Rate (ABR) ATM Transfer Capability | approved |
| Q.2723.4 | B-ISDN user part - Support of the ATM Block Transfer (ABT) ATM Transfer Capability | approved |
| Q.2723.5 | B-ISDN user part - Support of Cell Delay Variation Tolerance (CDVT) | not approved |
| Q.2723.6 | B-ISDN user part - Support of Global tagging | not approved |
| Q.2724.1 | B-ISDN user part - Look-ahead without state change for the Network Node Interface (NNI) | approved |
| Q.2725.1 | B-ISDN user part - Support of negotiation during connection setup | approved |
| Q.2725.2 | B-ISDN user part - Modification procedures | approved |
| Q.2725.3 | B-ISDN user part - Modification procedures for | approved |

| | | |
|---|---|---|
| | sustainable cell rate parameters | |
| Q.2725.4 | B-ISDN user part - Support of modification with negotiation | not approved |
| Q.2726.1 | B-ISDN user part - ATM end system address | approved |
| Q.2726.2 | B-ISDN user part - Call priority | approved |
| Q.2726.3 | B-ISDN user part - Network generated session identifier | approved |
| Q.2726.4 | B-ISDN user part - User generated identifiers | approved |
| Q.2727 | B-ISDN user part - Support of frame relay | approved |
| Q.27svp | B-ISDN user part - Switched Virtual Path Capability (SPVC) | not approved |
| Q.27spvc | B-ISDN user part - Soft permanent virtual path/channel connections (SVPC/SVCC) | not approved |

# Appendix B

# NOTATION FOR THE OBJECT MODELING TECHNIQUE

This appendix gives an overview of the notation for the Object Modeling Technique (OMT). However, no complete OMT description is provided, but only the notations are summarized which are used within the book. For more information about OMT please refer to [1].

It is obvious that the Unified Modeling Language (UML) [2] will become more important and seems to replace the OMT notation in the near future. Some of the notations used in the books already adopt UML aspects (e.g. the notation for object instances) or UML terminology (e.g. Interaction Diagram).

## B.1 Object Model Notation

'An *object model* captures the static structure of a system by showing the objects in the system, relationships between the objects, and the attributes and operations that characterize each class of objects'. [1] The Object Model can be described by two kinds of object diagrams: class diagrams and instance diagrams. A *class diagram* is a schema, pattern or template for describing a general abstraction of individual objects. It consists of classes and their relationships. An *instance diagram* is a description of a concrete situation, containing object instances of the corresponding class diagram.

Within this book, both kinds of object diagrams are used. Combination of class and instance diagrams in a single diagram, as suggested in [1], are not used here.

### *Notation for Class Diagrams*

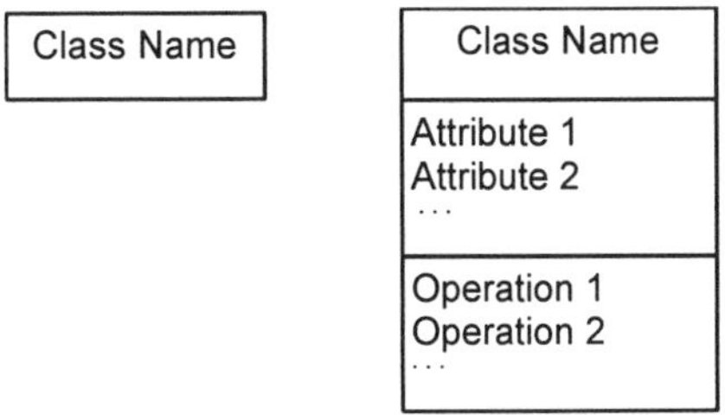


**Figure B.1:**
Notation for classes

*Intelligent Broadband Networks*. Edited by I. S. Venieris, H. Hussmann

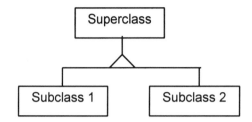

**Figure B.2:**
Notation for generalization (inheritance)

**Figure B.3:**
Notation for associations

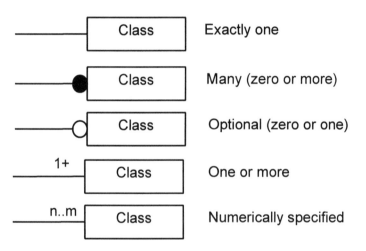

**Figure B.4:**
Notation for multiplicity of associations

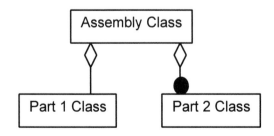

**Figure B.5:**
Notation for aggregation

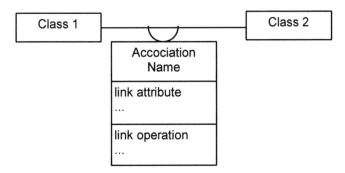

**Figure B.6**
Notation for association-as-class

## Notation for Instance Diagrams

To introduce optional object names or identifiers, a little deviation from this notation is used within this book, which is derived from the corresponding UML notation [2].

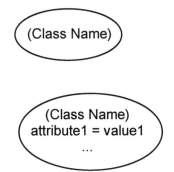

**Figure B.7**
Notation for object instances

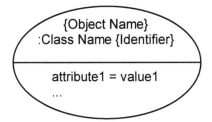

**Figure B.8:**
Extended notation for object instances

OMT provides no special notation for different relationships between objects.

Instead, it is referred to the relationship between the classes of which the objects are instances. However, to reflect the relationships between objects in instance diagrams, the same notation as for classes is used for aggregations and associations:

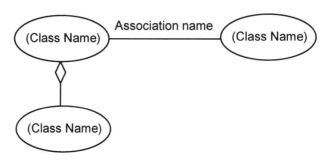

**Figure B.9:**
Notation for object relationships

## B.2    Dynamic Model Notation

In this book, two kinds of diagrams are used to model the dynamic behavior of the system: state diagrams and interaction diagrams. OMT makes intensive use of state diagrams but only mentions the need for interaction diagrams (which are called event trace diagrams in [1]). A formal notation for event trace diagrams is missing in OMT. Therefore for interaction diagrams a notation is used in this book which is derived from UML (see also [2], [3] and [4]).

*State Diagrams*

Only a small subset of the notation by OMT is used, which is presented in Figures B.9 and B.10.

**Figure B.10:**
Notation for states and transitions between states

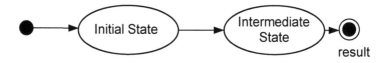

**Figure B.11:**
Notation for initial and final states

## *Interaction Diagrams*

Usually, the boxes contain the names of the instances (objects) and the messages sent from one instance to another is a method call of one object at the other. However, the notation of interaction diagrams is also used in a broader sense within this book by allowing arbitrary entities, e.g. functional units, network elements etc. instead of objects and any kind of communication between these entities.

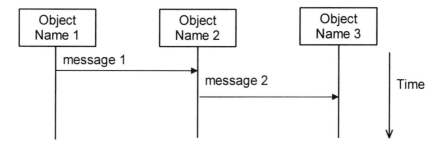

**Figure B.12:**
Notation for interacting objects/entities

# References

[1]  Rumbaugh J, Blaha M, Premerlani W, Eddy F and Lorensen W, *Object-Oriented Modeling and Design*, Prentice Hall, 1991

[2]  Rumbaugh J, Jacobson I and Booch G, *The Unified Modeling Language, Documentation Set 1.1*, Rational Software Corporation, Santa Clara, 1997

[3]  Booch G, *Object-Oriented Analysis and Design with Applications*, Second Edition, Benjamin/Cummings, 1994

[4]  Jacobson I, Christerson M, Jonsson P and Oevergaard G, *Object-Oriented Software Engineering - A Use-Case Driven Approach*, Addison-Wesley, 1992

# Appendix C

## THE M/G/1 QUEUING SYSTEM WITH PRIORITIES AND FEEDBACK

The M/G/1 queuing system with priorities and feedback is a tractable analytical tool that can be efficiently used for modeling and evaluating complex signaling protocols and IN functional entities. Each protocol is represented by a single queue, while the protocol functions constitute internal queues each one with a different priority level ($p$). In the following analysis two kinds of messages are assumed: messages that enter the system from outside and messages that are feed-backs of other messages. Messages ($i$) belonging to the first category define a message class ($s(i)$), containing themselves as well as all their feed-backs. The first message of each class is called the generator message of this class and it is the only message for which the relation $s(i)= i$ holds. For all messages $j$ that belong to the message class generated by the message $i$, the relation $s(j)=i$ holds. In addition, $j+1$ is the first feed-back of message $j$, if both belong to the same class ($s(j)=s(j+1)$).

A priority $p(j)$ is assigned to each message $j$ according to the particular protocol function (i.e. step in the internal itinerary) the message refers to. Higher $p(j)$ values declare higher priority. All service times $t_j$ are assumed deterministic and external arrivals are modeled as a Poisson process with rate $\lambda_j$. Feed-back messages arrive at rates equal to the arrival rate of the generator message of the message class, i.e. $\lambda_j = \lambda_{s(j)}$ for all messages belonging to the same class $s(j)$.

The probability that a $j$ message is being serviced can be derived easily by applying Little's theorem: $\rho_j = \lambda_j\, t_j$, while the stability condition of the system is obviously: $\sum_{j=1}^{N} \rho_j < 1$.

For the following analysis the analytical model described in [1, 2], is adopted with the difference that service priorities are not preemptive. This assumption simplifies the model but nevertheless it has serious impact in the form of the equations. The expected time that a message $j$ spends in the system, $W_j$, is defined as the time interval from the time the message entered the system until the time its service is completed. If $Q_j$ is the 'steady-state' expected number of $j$ messages in the system, then $Q_j = \lambda_j W_j$ or $Q = \Lambda W$ in matrix notation, where $\Lambda$ is a diagonal matrix with $\Lambda_{jj} = \lambda_j$ and $W$ is the matrix with elements the waiting times $W_j$. It is proved in [2] that a column vector $X=(x_1,x_2,...,x_N)^T$, where $x_j$ is the expected number of $j$ messages in the queue, can be used to characterize the state of the system. The message being serviced -if any- does not count.

Message types can be distinguished in two categories depending on the state of the system at the times they enter the system. The first category contains the generator messages, while the second the feed-back messages. This separation is imposed by the fact that generator messages may find a message being serviced at the moment they enter, while feed-back messages always find the server unoccupied.

The expected time in the system of a message $j$ that has just fed-back is the sum of three terms: the total time spent while waiting for messages with higher or equal priority already in the system at the moment of the message $j$ arrival, the time waiting for higher priority messages that arrive while the message $j$ waits in the queue and the service time of the message $j$. In the case of a generator message a fourth term must be added, declaring the time spent waiting for the message in the server to end its service. In analogy to [2] we can derive that:

$$W_j = \frac{W_j^{(1)} + \sum_{p(i) \geq p(j)} GD_j(i) - \sum_{p(i) \geq p(j), i = s(i)} \lambda_i t_i D_j(i) + t_j}{1 - \sum_{p(i) > p(j), i = s(i)} \lambda_i D_j(i)} \qquad \text{(C-1)}$$

but now $D_j(i)$, which is the expected total delay that a generator message $j$ suffers due to an $i$ message already in the system , is given by:

$$D_j(i) = \begin{cases} 0 & \text{if } p(i) < p(j) \\ t_i + L_j(i+1) & \text{if } p(i) \geq p(j) \end{cases} \quad \text{and}$$

$$L_j(i) = \begin{cases} 0 & \text{if } p(i) \leq p(j) \\ t_i + L_j(i+1) & \text{if } p(i) > p(j) \\ & \text{and } s(i-1) = s(i) \end{cases}$$

$W_j^{(1)}$ is the delay term due to the message inside the server and is given by:

$$W_j^{(1)} = \sum_{p(i) \geq p(j)} \rho_i (\tilde{t}_i + D_j(i+1) 1_{p(i+1) > p(j)}) , \text{ where } \tilde{t}_i \text{ is the expected time}$$

an $i$ message will remain under service as seen by a Poisson arrival process. $W_j^{(1)}$ is 0 in the case of a message entering the system as feed-back.

If $j \neq s(j)$, i.e. j is a feed back message, then the term $G$ in Eq. (C-1) is the random vector $V = (v_1, v_2, ..., v_N)^T$ of the number of messages in the system. In the case where $j = s(j)$ this term becomes $\lambda_i W_i - \rho_i$ , i.e. $G$ equals the state of the system.

Eq. (C-1) can be written in a simpler way if matrix notation is adopted:

$$W_j(X) = T_j X + r_j \qquad \text{if } j \neq s(j) \qquad \text{or} \qquad \text{(C-2a)}$$

$$W_j = T_j(\Lambda W - \rho) + r_j \quad \text{if } j = s(j) \qquad \text{(C-2b)}$$

where:
$$r_j = \frac{W_j^{(1)} - \sum_{p(i)>p(j),i=s(i)} \lambda_i t_j D_j(i) + t_j}{1 - \sum_{p(i)>p(j),i=g(i)} \lambda_i D_j(i)}$$
and

$$T_j(i) = \begin{cases} 1 - \dfrac{D_j(i)}{\sum_{p(i)>p(j),i=s(i)} \lambda_i D_j(i)} & \text{if } p(i) \geq p(j) \\ 0 & \text{otherwise} \end{cases}$$

Eq. (C-2a) stands for messages $j \neq s(j)$, while Eq. (C-2b) for generator messages $(j = s(j))$. Eqs (C-2a) and (C-2b) differ only in the state of the system that the entering message observes. When $j = s(j)$, the state of the system at the moment message $j$ enters is the steady state, i.e. $Q$-$\rho$, while this does not hold when $j \neq s(j)$. In the latter case, if message $j$ finds the system in state $X$ at the moment of its entering, then it can be proved [2] that the system state $Y$ when its service is completed will be:

$$Y = A_{p(j)} X + R_j \tag{C-3}$$

and, if $X_j^*$ is the state of the system when a type $j$ message enters the system, then:

$$X_{j_l}^* = (\prod_{i=1}^{l-1} A_{j_i})(Q-\rho) + (\prod_{i=2}^{l-1} A_{j_i}) R_{j_1}' + \sum_{k=2}^{l-1} (\prod_{s=k+1}^{l-1} A_{j_s}) R_{j_k} \tag{C-4}$$

In the above relation the parameter $j_l$ defines a chain in the message types, i.e. the itinerary defined by a generator message and all its feedback messages. Derivation of $A_{j_i}$, $R'_{j_l}$, $R_{j_k}$ can be found in [2].

From Eqs (C-2a), (C-2b) and (C-4) we can write the following formula:

$$W_j = T_j(\prod_{k=s(j)}^{j-1} A_k) \Lambda W - T_j(\prod_{k=s(j)}^{j-1} A_k)\rho + T_j(\prod_{k=s(j)+1}^{j-1} A_k) R'_{s(j)} + $$

$$T_j \sum_{s<j} (\prod_{s=k+1}^{l-1} A_k) R_s + r_j \tag{C-5}$$

Since Eq. (C-5) can give the waiting times for all the types of messages, we can derive easily relations giving other parameters of interest, such as the number of messages in each priority level and their mean waiting time, the average delay in the system, etc.

# References

[1] Paterok M, Fisher O, *Feedback Queues with Preemption Distances Priorities*, ACM Sigmetrics Performance Evaluation Review, **17**, 136-145, 1989
[2] Simon B, *Priority Queues with Feedback*, Journal of *ACM*, **31**, 134-149, 1984

# Appendix D

## SYSTEM PARAMETERS FOR PERFORMANCE EVALUATION

In this appendix, the system and traffic parameters used in the performance evaluation studies of this book are given. These include the processing times allocated to each function of a signaling or IN node, or alternatively the service time of each signaling or IN message in accordance with the particular method applied for modeling the system components, the call holding time and the priority scheme applied to schedule the sequence of service for each message.

### D.1   Processing Times

In both analytical and simulation studies appearing in Sections 4.3.1 and 4.5.1 constant service times per process have been assumed. The service times per process ($T$), within a functional entity or protocol, appear in Tables D.1-D.6 (the processing time is expressed in milliseconds).

The indicated processes correspond to the ones modeled with the network of queues approach in Section 4.2.2. The B-ISUP processing times were defined using as input the values of [1] for the time a message spends in the entire protocol. Taking into account the nature of the process activated into a specific protocol when a particular message arrives, the protocol time has been decomposed into the process times appearing in Tables D.1-D.6. The processing times for the Q.2931, CCF/SSF, SCF, SRF and SDF have been defined similarly taking into account the functions of each process. These values are used as constant service times when the analytic model of Appendix C is used (Section 4.3.1 and 4.5.1) and when simulations are performed (Section 4.5.1).

The tables also report the process priority (ps) obtained with the priority rules of Total and External Load defined in Section D.3. Also the utilization of each process is defined assuming a total rate of incoming calls equal to $\lambda$. By letting $M$ = number of messages per protocol process or functional entity and $E$ = number of external messages per protocol process or functional entity and assuming $\lambda_{VoD}$ = 0.15* $\lambda$ for VoD and $\lambda_{Voice}$ = 0.85*$\lambda$ for telephony, it follows that for each process $j$ we have:

- A utilization due to external arrivals $\rho'_j = (E_{VoD} *\lambda_{VoD})/\mu_j + (E_{Voice}*\lambda_{Voice})/\mu_j$ where the two terms represent the contribution of VoD and telephony respectively;
- A total utilization $\rho''_j = (M_{VoD} *\lambda_{VoD})/\mu_j + (M_{Voice}*\lambda_{Voice})/\mu_j$ where the two terms represent the utilization due to VoD and telephony respectively.

*Intelligent Broadband Networks.* Edited by I. S. Venieris, H. Hussmann
© 1998 John Wiley & Sons Ltd.

**Table D.1**
Parameters of Q2931 protocol

| Q2931 | $T$ ms | $\rho''$ | $\rho'$ | ps |
|---|---|---|---|---|
| Incoming user side | 0.15 | $0.36\lambda + 0.51\lambda$ | $0.36\lambda + 0.51\lambda$ | 2 |
| Outgoing user side | 0.25 | $0.56\lambda + 0.85\lambda$ | - | 1 |
| Incoming exchange | 0.15 | $0.135\lambda + 0.51\lambda$ | $0.135\lambda + 0.51\lambda$ | 2 |
| Outgoing exchange | 0.25 | $0.187\lambda + 0.85\lambda$ | - | 1 |

**Table D.2**
Parameters of B-ISUP protocol

| BISUP user side | $T$ ms | $\rho''$ | $\rho'$ | ps |
|---|---|---|---|---|
| SACF | 0.3 | $0.72\lambda + 2.55\lambda$ | $0.18\lambda + 0.765\lambda$ | 4 |
| BCC | 1.0 | $1.2\lambda + 1.7\lambda$ | - | 3 |
| CC | 0.5 | $0.45\lambda + 1.275\lambda$ | - | 2 |
| AP | 0.5 | $0.45\lambda + 1.7\lambda$ | $0.15\lambda + 0.425\lambda$ | 1 |

**Table D.3**
Parameters of B-ISUP protocol

| BISUP exchange | $T$ ms | $\rho''$ | $\rho'$ | ps |
|---|---|---|---|---|
| SACF | 0.3 | $0.72\lambda + 2.55\lambda$ | $0.09\lambda + 0.255\lambda$ | 4 |
| BCC | 1.0 | $0.6\lambda + 1.7\lambda$ | - | 3 |
| CC | 0.5 | $0.3\lambda + 1.275\lambda$ | - | 2 |
| AP | 0.5 | $0.45\lambda + 1.7\lambda$ | $0.15\lambda + 0.85\lambda$ | 1 |

**Table D.4**
Parameters of CCF/SSF IN functional entities

| CCF/SSF | $T$ ms | $\rho''$ | $\rho'$ | ps |
|---|---|---|---|---|
| BCM | 0.5 | $2.25\lambda$ | $0.975\lambda$ | 3 |
| FIM | 1.0 | $1.95\lambda$ | - | 2 |
| IN-SM | 1.0 | $1.95\lambda$ | $1.2\lambda$ | 1 |

**Table D.5**
Parameters of SCF IN functional entity

| SCF | $T$ ms | $\rho''$ | $\rho'$ | ps |
|---|---|---|---|---|
| FEAM | 0.3 | $1.8\lambda$ | $0.95\lambda$ | 1 |
| SLEM | 1.5 | $6.07\lambda$ | - | 2 |
| DAM | 0.5 | $0.075\lambda$ | - | 5 |
| SLP_L | 0.5 | $0.075\lambda$ | - | 3 |
| FRL | 0.5 | $0.075\lambda$ | - | 4 |

**Table D.6**
Parameters of SRF, SDF IN functional entities

| SRF | $T$ ms | $\rho''$ | $\rho'$ | ps | SDF | $T$ | $\rho''$ | $\rho'$ | ps |
|---|---|---|---|---|---|---|---|---|---|
| FEAM | 0.3 | $0.45\lambda$ | $0.225\lambda$ | 1 | FEAM | 0.3 | $0.63\lambda$ | $0.27\lambda$ | 1 |
| RM | 0.5 | $0.375\lambda$ | - | 2 | DM | 0.5 | $0.45\lambda$ | - | 2 |

In the single queue approach used in Chapter 4.4 all messages have an established *weight* which equals the number of mandatory information elements included in the message. The *weight* values are indicated in Tables D.7 to D.10 for a VoD service. Each message instance, relevant to the VoD information flow, is identified by its name, signaling channel and occurrence. For each phase, the occurrence represents the number of times each message appears in the same signaling channel.

**Table D.7**
Phase1: the STB is connected to the B-IP

| Message | Signaling Channel | Occurrence | Weight |
|---|---|---|---|
| Setup | STB→SSP | 1 | 6 |
| Service request | SSP→SCP | 1 | 14 |
| Request report SSM change | SCP→SSP | 1 | 9 |
| Join party to session and link leg to bearer | SCP→SSP | 1 | 5 |
| Setup | SSP→IP | 1 | 6 |
| Connect | IP→SSP | 1 | 2 |
| Connect | SSP→STB | 1 | 2 |
| Conn ack | SSP→IP | 1 | 1 |
| Report SSM change | SSP→SCP | 1 | 6 |
| Conn ack | STB→SSP | 1 | 1 |
| Request report SSM change | SCP→SSP | 2 | 9 |
| Add bearer to session | SCP→SSP | 1 | 8 |
| Setup | SSP→STB | 1 | 6 |
| Setup | SSP→IP | 2 | 6 |
| Connect | STB→SSP | 1 | 2 |
| Connect | IP→SSP | 2 | 2 |
| Report SSM change | SSP→SCP | 2 | 6 |
| Conn ack | SSP→STB | 1 | 1 |
| Conn ack | SSP→IP | 2 | 1 |
| Play announcement | SCP→IP | 1 | 3 |
| Sr report | IP→SCP | 1 | 3 |
| Prompt & c | SCP→IP | 1 | 3 |
| Collected info | IP→SCP | 1 | 4 |
| Play announcement | SCP→IP | 2 | 3 |
| Sr report | IP→SCP | 2 | 1 |
| Prompt & c | SCP→IP | 2 | 3 |
| Collected info | IP→SCP | 2 | 6 |
| Prompt & c | SCP→IP | 3 | 3 |
| Collected info | IP→SCP | 3 | 4 |

**Table D.8**
Phase2: the STB is disconnected from the B-IP

| Message | Signaling Channel | Occurrence | Weight |
|---|---|---|---|
| Drop party | SCP→SSP | 1 | 3 |
| Release | SSP→STB | 1 | 2 |
| Release | SSP→IP | 1 | 2 |

| Report SSM change | SSP→SCP | 1 | 6 |
|---|---|---|---|
| Rel comp | STB→SSP | 1 | 1 |
| Rel comp | IP→SSP | 1 | 1 |
| Release | SSP→STB | 2 | 2 |
| Release | SSP→IP | 2 | 2 |
| Report SSM change | SSP→SCP | 2 | 8 |
| Rel compl | STB→SSP | 2 | 1 |
| Rel compl | IP→SSP | 2 | 1 |
| Furnish charging info | SCP→SSP | 1 | 2 |

**Table D.9**
Phase3: the STB is connected to the VS

| Message | Signaling Channel | Occurrence | Weight |
|---|---|---|---|
| Request report SSM change | SCP→SSP | 1 | 9 |
| Add party and bearer to session | SCP→SSP | 1 | 10 |
| Setup | SSP→STB | 1 | 6 |
| Setup | SSP→VS | 1 | 6 |
| Connect | STB→SSP | 1 | 2 |
| Connect | VS→SSP | 1 | 2 |
| Report SSM change | SSP→SCP | 1 | 6 |
| Conn ack | SSP→STB | 1 | 1 |
| Conn ack | SSP→VS | 1 | 1 |
| Request report SSM change | SCP→SSP | 2 | 9 |
| Add bearer to session | SCP→SSP | 1 | 8 |
| Setup | SSP→STB | 2 | 6 |
| Setup | SSP→VS | 2 | 6 |
| Connect | STB→SSP | 2 | 2 |
| Connect | VS→SSP | 2 | 2 |
| Report SSM change | SSP→SCP | 2 | 6 |
| Conn ack | SSP→STB | 2 | 1 |
| Conn ack | SSP→VS | 2 | 1 |

**Table D.10**
Phase4: the STB is disconnected from the VS

| Message | Signaling Channel | Occurrence | Weight |
|---|---|---|---|
| Release | STB→SSP | 1 | 2 |
| Release | SSP→VS | 1 | 2 |
| Report SSM change | SSP→SCP | 1 | 6 |
| Rel compl | SSP→STB | 1 | 1 |
| Rel compl | VS→SSP | 1 | 1 |
| Release session | SCP→SSP | 1 | 1 |
| Release | SSP→STB | 1 | 2 |
| Release | SSP→VS | 2 | 2 |
| Report SSM change | SSP→SCP | 2 | 6 |
| Rel compl | STB→SSP | 1 | 1 |
| Rel compl | VS→SSP | 2 | 1 |

The processing times per message appearing in Tables D.11 to D.13 have been used in the simulation study of Section 4.5.2 and have been obtained experimentally from the prototype platform described in Chapter 2.8. All values are expressed in Time Units, where the Time Unit is assumed to be the processing time of the Service Request message in the B-SCP. The reason for not using absolute processing time values as done with Tables D.1-D.6 is that the assumption of identical B-SSPs holding for the theoretical values is not valid in the prototype configuration where the participating B-SSPs were originated from different manufacturers. Measurements have been performed for a B-VPN and plain telephony service. The B-ISUP message processing times (in Time Units) are estimated from measurements on the Q.2931 protocol.

**Table D.11**
The service time ratio of the B-INAP messages processed by the B-SSP and B-SCP physical entities, when a B-VPN service is used

| Entity | Incoming Message | Outgoing Message | Service time [Time Units] |
|---|---|---|---|
| B-SSP | 3*Request report SSM change + Join party to session and Link leg to bearer | Report SSM change | 0.29381 |
| | Continue | Call proc | 0.13745 |
| | Continue | Setup | 0.22164 |
| | Release session | OUT(end) | 0.13745 |
| B-SCP | Service request | 3*Request report SSM change + Join party to session and Link leg to bearer | 1 |
| | Report SSM change | Continue | 0.27319 |
| | Report SSM change | Release session | 0.39003 |
| | Report SSM change | OUT(end) | 0.27319 |

**Table D.12**
The service time ratio of the Q.2931 and B-ISUP messages processed by the B-SSP and User physical entities, when a B-VPN service is used

| Entity | Incoming Message | Outgoing Message | Service time [Time Units] |
|---|---|---|---|
| B-SSP | Setup | Service request | 0.40034 |
| | Connect | Report SSM change | 0.39690 |
| | Connect | Connect ack | 0.44158 |
| | Connect | Connect | 0.45704 |
| | Call proc | OUT(end) | 0.04467 |
| | ANM | CONNECT | 0.31958 |
| | IAM | IAA | 0.20790 |
| | IAM | IAM | 0.29037 |
| | IAM | Setup | 0.29037 |
| | Connect | Connect ack(in this case Report | 0.30584 |

|      |               | SSM change is not created) |         |
|------|---------------|----------------------------|---------|
|      | Connect       | Connect (in this case Report SSM change is not created) | 0.31958 |
|      | IAA           | OUT(end)                   | 0.07560 |
|      | Release       | Release                    | 0.02920 |
|      | Release       | Report SSM change          | 0.20790 |
|      | Release       | Release compl              | 0.30068 |
|      | REL           | REL                        | 0.02920 |
|      | REL           | Release                    | 0.02920 |
|      | REL           | RLC                        | 0.30068 |
|      | Release compl | OUT(end)                   | 0.02920 |
| User | Setup         | Call proc                  | 0.11168 |
|      | Setup         | Connect                    | 0.89347 |
|      | Connect       | Connect ack                | 0.07560 |
|      | Release       | Release compl              | 0.86254 |

**Table D.13**
The service time ratio of the Q.2931 and B-ISUP messages processed by the B-SSP and User, physical entities, when a telephony service is used

| Entity | Incoming Message | Outgoing Message | Service time [Time Units] |
|--------|------------------|------------------|---------------------------|
| B-SSP  | Setup            | Call proc        | 0.20790 |
|        | Setup            | Setup            | 0.29037 |
|        | Connect          | Connect ack      | 0.30584 |
|        | Connect          | Connect          | 0.31958 |
|        | Call proc        | OUT(end)         | 0.01374 |
|        | IAM              | IAA              | 0.20790 |
|        | IAM              | IAM              | 0.29037 |
|        | IAM              | Setup            | 0.29037 |
|        | IAM              | ANM              | 0.29037 |
|        | IAA              | OUT(end)         | 0.07560 |
|        | Release          | Release          | 0.02920 |
|        | Release          | Release compl    | 0.19759 |
|        | REL              | REL              | 0.02920 |
|        | REL              | Release          | 0.02920 |
|        | REL              | RLC              | 0.19759 |
|        | Release compl    | OUT(end)         | 0.02920 |
| User   | Setup            | Call proc        | 0.11340 |
|        | Setup            | Connect          | 0.92611 |
|        | Connect          | Connect ack      | 0.04982 |
|        | Release          | Release compl    | 0.75429 |

## D.2    Call Duration

For the VoD service, movies with a mean length of 110 minutes [3] are assumed. Users either watch the whole movie or stop the movie 5 minutes after the selection. The VoD duration ($d$) in the simulation follows a distribution with $d$=110 min with probability $p_1$=0.85 and $d$=5 min with probability $p_2$=0.15. The duration of telephony follows a mixture of two normal distributions on logarithmic time scale, that is $F(t) = \beta \, F_1(t) + (1-\beta) \, F_2(t)$. Realistic values for the parameters of the previous distributions are given in [3]: mean call duration 150 sec, $\beta = 0.4$, $\mu_1 = 1.31$, $\sigma_1 = 0.33$, $\mu_2 = 2.11$, $\sigma_2 = 0.5$.

The same distribution is used for the B-VPN with smaller values of the parameters to accelerate the simulation process : $\beta = 0.4$, $\mu_1 = 0.655$, $\sigma_1 = 0.165$, $\mu_2 = 1.055$, $\sigma_2 = 0.25$. In all cases the call duration is much higher than the call set up delay, so that the accuracy of results is not influenced [4].

## D.3    Priority Schemes

A further characteristic of the utilized model is the priority scheme used to serve the queues representing the processes activated in a protocol or in a functional entity. In the following the two priority assignment schemes used in Chapters 4.3 and 4.5 of this book are presented.

### Rule of Total and External Load

In this scheme priorities are assigned in accordance to the following steps (the utilizations $\rho'$ and $\rho''$ used here follow the definitions given D.1):

- Firstly the process with the higher priority must be singled out. This could be accomplished by applying the following rules: the internal process with the higher utilization ($\rho''$) that is crossed by multiple itineraries is always allocated the higher priority. This rule is called the Rule of Total Load (RTL). If there is no process with the above features two cases are distinguished: if there is more than one process that accepts external messages the process with the lower utilization ($\rho'$) due to external arrivals is allocated the higher priority. This rule is called the Rule of External Load (REL). If there is only one process that accepts external messages the process with the lower utilization ($\rho''$) is allocated the higher priority.
- The priority of the other processes are assigned by considering all the internal itineraries starting from the process selected with the above rules and by applying the following criteria: if the itinerary is only one, the priority is assigned according to the order the processes appear in the itinerary. If there is a multiplicity of itineraries starting from the selected process, the higher priority is assigned to the process with lower utilization ($\rho''$). This is accomplished in a recursive manner.

The rationale under the proposed criteria is to limit the delay experienced by the messages that cross the network elements (protocols or functional entities).

### The Time-Based priority scheduling

By using this priority scheme, new and cycled messages belonging to the same IN request acquire the same priority. The priority is assigned based on the arrival time of the request.

In the Time Based priority scheduling algorithm (see Figure D.1) an earlier message to the system will get a higher priority than any message that arrived into the system on a later moment. Let messages be ordered according to their arrival times to the network. Then the $m$-th message is assigned priority $PR(m)$:

$$PR(m) = \begin{cases} MAXIMUM, & m = 0 \\ PR(m-1)-1, & m > 0 \end{cases}$$

whereby, $MAXIMUM$, is the highest possible priority that a message can get.

Suppose that a message $m1$ arrives at the B-SSP at time step $t1$ and acquires priority $PR1$. This message will be served by the B-SSP and then sent to the B-SCP. Let's assume that a new message $m2$ arrives at the B-SSP at time step $t2$ $(t2 > t1)$ and gets priority $PR2$ where $PR2 < PR1$. Suppose that message $m2$ will start to be served by the SSP and before completion of its service message $m1$ re-enters the B-SSP. Message $m1$, which has been assigned a higher priority than $m2$, will pre-empt $m2$ from the server. After $m1$ departs from the B-SSP, message $m2$ will continue to be serviced.

# References

[1]   Veeraraghavan M, La Porta T, Lai W, *An Alterantive Approach to Call/Connection Control in Broadband Switching Systems*, IEEE Commun. Mag., **33**, 90-96, 1995

[2]   Ghafir H., Chadwich H., *Multimedia Servers - Design and Performance*, Procceedings of Globecom '94, 1994

[3]   Bolotin V, *Modeling Call Holding Time Distributions for CCS Network Design and Performance Analysis*, IEEE J. Select. Areas Commun., **12**, 433-438, 1994

[4]   Veldkamp E.P., *Performance Aspects of the Platinum Signaling System*, CTIT Technical Report series, 96-98, 1996

# Index

*Intelligent Broadband Networks.* Edited by I. S. Venieris, H. Hussmann
© 1998 John Wiley & Sons Ltd.